INTERNATIONAL YEARBOOK OF LIBRARY AND
INFORMATION MANAGEMENT 2002/2003

The Digital Factor in Library and Information Services

Edited by

G. E. Gorman

*School of Information Management,
Victoria University of Wellington*

THE SCARECROW PRESS, INC.
LANHAM, MARYLAND, 2002

SCARECROW PRESS, INC.

Published in the United States of America
By Scarecrow Press, Inc.
4720 Boston Way
Lanham, Maryland 20706

http://www.scarecrowpress.com

This edition published simultaneously in Lanham, Maryland, by Scarecrow
Press, and in London, UK, by Facet Publishing, 2002.

ISBN 0-8108-4590-3

Library of Congress Cataloguing-in-Publication Data Available

Typeset from editor's disks by Facet Publishing in 11/15pt New Baskerville
and Franklin Gothic Condensed.
Printed and made in Great Britain by MPG Books Ltd, Bodmin, Cornwall.

To the information community in Lebanon,
working to provide quality service in a
difficult but hopeful environment

Contents

Editorial Advisory Board 2002/2003

Australia and New Zealand

- **Dr Peter Clayton**, Faculty of Communication and Education, University of Canberra, Canberra 2614, Australia.
  E-mail: prc@comedu.canberra.edu.au
- **Ms Rowena Cullen**, School of Information Management, Victoria University of Wellington, PO Box 600, Wellington 6015, New Zealand.
  E-mail: rowena.cullen@vuw.ac.nz

China and Vietnam

- **Dr Van Son Vu**, National Centre for Scientific and Technological Information and Documentation, 24 Ly Thuong Kiet Street, Hanoi, Vietnam.
  E-mail: vson@vista.gov.vn
- **Dr Jianzhong Wu**, The Shanghai Library, 1555 Huai Hai Zhong Lu, Shanghai 200031, China.
  E-mail: jzwu@libnet.sh.cn

South Asia

- **Dr M. P. Satija**, Department of Library and Information Science, Guru Nanak Dev University, Amritsar 143005, India.
  E-mail: dcse_gndu@yahoo.com

Eastern Africa

- **Mr Stephen M. Mutula**, Department of Library and Information Studies, University of Botswana, Private Bag 0022, Gaborone, Botswana.
  E-mail: mutulasm@mopipi.ub.bw

United Kingdom

- **Professor Peter Brophy**, Department of Information and Communications, Manchester Metropolitan University, Geoffrey Manton Building, Rosamond Street West, off Oxford Road, Manchester M15 6LL, England.
  E-mail: p.brophy@mmu.ac.uk

USA and Canada

- **Ms Peggy Johnson**, University Libraries, 499 Wilson Library, University of Minnesota, 309 19th Avenue South, Minneapolis, MN 55455, USA.
  E-mail: m-john@maroon.tc.umn.edu
- **Dr Lorna Peterson**, University at Buffalo, State University of New York, NY14260, USA.
  E-mail: lpeterso@acsu.buffalo.edu

Western Europe

- **Mr Paul S. Ulrich**, Information Services, Zentral- und Landes-

About the contributors

Dr Denice Adkins is Assistant Professor in the School of Information Science and Learning Technologies at the University of Missouri, USA. She has worked in reference and children's services in US public libraries.

Professor Peter Brophy is Professor of Information Management and Director of the Centre for Research in Library and Information Management (CERLIM) in the Department of Information and Communications at the Manchester Metropolitan University, UK.

Dr Christopher Brown-Syed is currently Visiting Assistant Professor at the School of Informatics, University at Buffalo, State University of New York, and Editor of the journal *Library and Archival Security*. He previously worked for the library system vendors Geac and Plessey.

Judith Clark is Head of Academic Support Services at the University of Ballarat, Australia. Previously she was the Manager of Information Services at the Cairns Campus of James Cook University in Queensland. Most recently, she has worked on a range of projects designed to improve the use of technology in teaching and learning, particularly in relation to the use of web-accessible information services and resources.

David Dawson is Senior ICT Adviser at Resource: The Council for Museums, Archives and Libraries in London, and UK Co-ordinator of the EU Digitising Content Together initiative. He was previously Outreach Manager (ICT) at mda, and Curator of the County Museum and Head of Documentation at Oxfordshire Museums Service.

Dr Marilyn Deegan is Director of Forced Migration Online at the Refugee Studies Centre at Oxford University, and was instrumental in the establishment of the Oxford Digital Library. She is Editor of *Literary and Linguistic Computing*, the Journal of the Association for Literary and Linguistic Computing, published by Oxford University Press, and Co-Director of the Office for Humanities Communication based at King's College London. With Simon Tanner she is co-author of a work closely related to her chapter in the Yearbook: *Digital futures: strategies for the information age* (Library Association Publishing, 2002).

Dr Daniel G. Dorner is Senior Lecturer and MLIS Programme Director in the School of Information Management at Victoria University of Wellington, New Zealand. He is also a contributor to the inaugural volume of the *International yearbook of library and information management* and has been involved in various metadata-related research projects.

Dr G. E. Gorman is Professor of Library and Information Management in the School of Information Management at Victoria University of Wellington, New Zealand. He is the founding general editor of the *International yearbook of library and information management* and the author or co-author of more than a dozen books and more than 100 refereed journal articles. He is also editor of *Online Information Review* (Emerald), Associate Editor of *Library Collections, Acquisitions and Technical Services* (Elsevier) and a member of the editorial boards of several other journals.

Shirley Hyatt is currently Communications and Business Transitions Director of the Office of Research, OCLC Online Computer Library Center, Inc, USA. Her responsibilities include recognizing the potential of technologies and innovations and ushering them into OCLC's development and marketing environments. Prior to joining OCLC's Office of Research, Ms Hyatt served as Director of Distributed Systems and Manager of OCLC's Access Services product line.

Shadrack Katuu is a lecturer in the Department of Library and Information Studies at the University of Botswana. His research interests include network security, record authentication and the management of electronic records. His work has been published in *Information for Development, ESARBICA Journal, Archivi per la Storia* and elsewhere.

Diane Kresh is Director of the Collaborative Digital Reference Service (now called QuestionPoint), USA, a project to build a global, web-based reference service among libraries and research institutions. She is a frequent speaker at professional meetings and conferences and the author of several articles on digital reference services. Ms Kresh is the recipient of the 2001 Federal 100 for her role in launching the Collaborative Digital Reference Service.

Stephen M. Mutula is a lecturer in the Department of Library and Information Studies, University of Botswana, where he teaches primarily in the areas of digital libraries and online information. He has previously worked in Kenya and Djibouti and has published extensively in the areas of internet connectivity, the use of IT in libraries and information centres, and the digital divide.

Dr Lorna Peterson is Associate Professor, Library and Information Studies, at the School of Informatics, University at Buffalo, State University of New York.

Dr Catherine Sheldrick Ross is Professor and Dean of the Faculty of Information and Media Studies at the University of Western Ontario, Canada. The research for her chapter was supported by a grant from the Social Sciences and Humanities Research Council of Canada. Dr

Ross has published a number of books, including four non-fiction books for children, a literary biography of Alice Munro and two books for library professionals (*Communicating professionally*, 2nd edn, Neal-Schuman and Library Association Publishing, 1998; *Conducting the reference interview*, Neal-Schuman and Facet Publishing, 2002). She teaches in the postgraduate programmes in library and information science at the University of Western Ontario. She is also engaged in a long-term research project, based on over 200 qualitative interviews with avid readers, on reading for pleasure.

Alastair G. Smith is Senior Lecturer in the School of Information Management at Victoria University of Wellington, New Zealand. In the 1980s he was involved in the development of online information retrieval systems at the Building Research Association of New Zealand and at the National Library of New Zealand. At VUW his research interests include methods for the evaluation of world wide web information resources.

Dr Sherry Shiuan Su is Associate Professor in the Department of Library and Information Science, Fu-Jen Catholic University, Taipei, Taiwan. Her research interests include information-seeking behaviour in the electronic environment and the impact of the internet on library services. She is currently working on a project about virtual reference interviews.

Simon Tanner is Senior Consultant for the Higher Education Digitisation Service, University of Hertfordshire. He was formerly Head of Library Services at Rolls-Royce and Associates, and has held information and IT posts at IBM and the Pilkington Library, Loughborough University. His consultancy activities have centred on digitization, project management and digital library development for clients such as Oxford University and the British Library. With Dr Marilyn Deegan he is co-author of *Digital futures: strategies for the information age* (Library Association Publishing, 2002).

Introduction

Each year the *International yearbook of library and information management* fixes on a specific theme which its contributors address from a variety of perspectives. The first two volumes focused respectively on collection management and information services, and in both instances the digital factor emerged as a common sub-theme. In a sense, then, the principal theme for this third volume was self-selected, and *The digital factor in library and information services* looks at how digital initiatives are affecting institutional models and finance, the packaging of information for different types of users, reference services, networks, collection management, security and a host of other issues important in the provision and management of library services.

Part 1

This year we open the collection in Part 1 with a statement cum query: In Praise of the Digital Revolution? For the first chapter, 'The spectrum of digital objects in the library and beyond', Dr Marilyn Deegan of Oxford University was asked to provide general background to digital library developments and to describe some of the key underlying principles (such as what digital libraries are and why they are important). Dr Deegan has performed her task to a high standard, and the result is a clear and insightful introduction that examines the background to digital library developments, looking at the nature of digital objects, the changing nature of libraries as organizations, and some of the international issues in collaboration and access which can both help and hinder the wide uptake of digital library developments. This chapter is largely in praise of the digital library; a different fix is taken by Dr Lorna Peterson in Chapter 2, 'Digital versus print issues'. She offers a synthesis of the digital and print issues currently exercising the minds of information professionals, and presents positive and negative features of both formats. Peterson concludes that the digital revolution will continue apace and that there will be plenty to 'amuse' those who are addicted to the newest technologies. At the same time, there is a continuing need for, and interest in, traditional formats. We should not dismiss proponents of one as geeks or of the other as Luddites – there is room and need for both in the unfolding information environment.

Part 2

Part 2, Institutional Models and Finance, opens with Professor Peter Brophy's 'New models of the library in a digital era'. In response to the question 'Do we really need traditional libraries?', Brophy suggests that we need to understand what libraries are and what they can be in the future. One way to achieve this understanding is through 'well-articulated and robust models' – he offers a range of such models in his provocative chapter. Brophy concludes that it is important for us to develop integrative models that help plan a secure future for library services in a world of rapid change. Peter Brophy has been a member of the *IYLIM* Editorial Board since its inception and now wishes to stand down. It is therefore appropriate that we have such a thoughtful chapter from him at this time. Thank you, Peter, for contributing to the success of the *Yearbook* since its inception – your insights have been especially valuable, and your services will be missed.

The second chapter devoted to institutional analysis is Simon Tanner's 'The economic opportunities and costs of developing digi-

tal libraries and resources'. This explores the core strategic and economic dilemmas faced by librarians in purchasing digital resources
and developing digital libraries. Tanner maintains that the opportunity cost of digital libraries is large, and with finite, often reducing,
budgets the librarian has never faced so many complex and important economic choices. This chapter describes these choices faced by
all librarians with regard to providing digital information resources,
shows how some libraries have reacted and what financial and management strategies are available to cope with the implementation of
new technologies and the provision of digital information.

Part 3

Part 3, 'Books and "Readers"', contains three chapters, opening with
what some might regard as an anachronistic topic in the 'digital bite'
era, a discussion of 'Reading in the digital age' by Professor Catherine
Sheldrick Ross of the University of Western Ontario. But is it really
so anachronistic? As Ross herself says in the opening lines of this
chapter, 'At a time when digital forms of information and entertainment are gaining in importance, there is considerable worried talk
about the death of reading and the obsolescence of the physical
book'. But, she maintains, this is a recurring anxiety that appears with
every new means of communication. What we really require is a dis-
passionate, evidence-based discussion of how readers actually engage
with different formats of digital text as distinct from printed text, why
they choose one format over another, and the different values and
satisfactions they assign to reading each. This is precisely what her
chapter seeks to achieve, and it is therefore integral to any discussion
of the digital factor in library and information services.

Chapter 6, by OCLC's Shirley Hyatt, is 'Judging a book by its cover:
e-books, digitization and print on demand'. This examines reading
ecologies, through an analysis of three sets of trends: in authoring, publishing and bookselling; in printing, finishing and distribution; in content, reading and e-books. She concludes that in order for the e-book

to become integrated into our lives and libraries (and this surely will occur) we need effective, efficient, human-oriented search software that will enable us to locate texts quickly and with ease; website designs that enable us to find what we are looking for quickly; portals that are reliable and well maintained; improved metadata; unique identifiers to distinguish one object from another; and credibility mechanisms. These may be future developments, but in Hyatt's view the future is not far away, and this future clearly bears considerable relation to Professor Ross's views on reading and our engagement with text in various formats.

The final chapter in Part 3 looks at users of web-based library services, but from a much-neglected perspective – 'The digital library and younger users'. Here Dr Denice Adkins of the University of Missouri reminds us that children and young adults have specific needs, and a digital library designed for adults will not serve those needs very well. Younger users want entertainment, collaboration and convenience with their information. Some of the barriers in providing digital library service to children are lack of computer and internet access, inability to use text-oriented search methods, and a lack of materials presented at a developmentally appropriate level. Adkins reminds us that digital libraries share a common history with traditional libraries, and today digital libraries for younger users are working to overcome barriers to youth access in much the same way that traditional libraries did in decades past. There is clearly a stimulating future for digital libraries aimed at younger users, and this chapter points the direction for appropriate development.

Part 4

'Reference services' contains four chapters, making it the most substantial section in the collection – appropriately so, for it is in the services to users that digitization is having the greatest impact on libraries, which after all remain service organizations, albeit in an increasingly competitive environment. And in a sense this part is a continuation of the discussion begun in the 2000–2001 *Yearbook*, which addressed information services generally, and included considerable discussion of digital factors in reference service.

Chapter 8, 'Web-based reference services: design and implementation decisions', is the first of two contributions looking at the impact of the web on reference services. In it Stephen Mutula of the University of Botswana discusses the planning, design and implementation of web-based reference services from the organizational perspective. His chapter is very wide ranging, moving from the evaluation and selection of web materials, to their organization and management, web-based services (especially remote reference service), training needs and design issues (including content and structure).

This discussion is ably complemented by Dr Sherry Shiuan Su in Chapter 9, 'Web-based reference services: the user-intermediary interaction perspective'. There is of necessity a small amount of overlap with Mutula's chapter, but overall Su takes a rather different perspective, looking not at the technology but the human interaction perspective. Human interaction in web-based reference work is discussed less frequently in the literature, and the purpose of this chapter is to explore the problems and issues concerning the characteristics of web-based reference service from the user-intermediary interaction perspective. Su reminds us that websites and web-based reference services share a similar goal – to provide information and services to users. Accordingly, understanding how users seek information on websites and interact with intermediaries is relevant to the design of web-based reference services. We must not forget that intermediaries, as a link between users and information, need to play a crucial role in helping users express their information needs.

In Chapter 10 Judith Clark discusses 'Digital library initiatives for academic teaching and learning: towards a managed information environment for online learning'. The concept of a managed information environment for online learning bridges the gap between the two domains of 'learning' (pedagogically guided) and 'library' (unguided reference) resources. This chapter looks at some of the ways in which academic libraries are working towards closer integration of library services and online learning systems. It argues that both learning domains need to be considered in rethinking library services to support flexible delivery in universities. A variety of approaches is considered, and the chapter concludes that libraries need a better understanding of the needs of online learners in regard to their use of electronic resources.

The final chapter in Part 4 is Diane Kresh's 'It's just a click away, or is it? The challenge of implementing a global information network'. As one who spends considerable time working with groups in less developed countries as diverse as Vietnam and Lebanon, I am acutely aware that the information 'superhighway' is as unreal as the Yellow Brick Road for many people – and this is the starting point for Kresh's chapter: 'for more than 80% of the world, the web is nowhere, its accessibility a remote concept that falls to the bottom of the hierarchy of needs of daily life'. We need to know who really uses the web, and the nature of the digital divide that many in the developed north dismiss as irrelevant (a topic ably tackled by my colleague Rowena Cullen in 'Addressing the digital divide', *Online Informaton Review*, **25** (5), 311–20), how the divide can be bridged and the role of libraries as 'bridges'. All of this is discussed by Ms Kresh, who sees the solution to the digital divide in the globalization of information and information services such as the Library of Congress's QuestionPoint, which operate above spatial and social limitations.

Part 5

Part 5 is devoted to Collection Management, or at least to the aspects of collection evaluation and retrospective digitization. The section opens with Alastair Smith's 'Evaluating digital collections', a topic that is arousing considerable interest among managers as digital collections become increasingly important in the overall service provision of libraries. His chapter reviews some evaluation methods proposed in the literature and proposes a checklist of criteria that can be used in relation to digital collections. The resulting Evaluation Checklist for Digital Collections will go some way towards establishing a standard for evaluating digital resources in libraries. Smith also discusses such related issues as evaluation of full-text services, databases and the web.

In Chapter 13 ('Creating content together: an international perspective on digitization programmes') David Dawson of Britain's Resource: The Council for Museums, Archives and Libraries writes on initiatives to create digital content – a field that is 'rich in assertions, and poor in proven experience'. His chapter aims to outline some of the key developments in this area and to identify some of the lessons that have been learned during developments in the UK and elsewhere in Europe in particular. In general we seem to be moving from small-scale experiments to large-scale, well-funded projects aimed at digitizing vast amounts of material. This is bound to have a significant impact on the future of library and information services, undoubtedly tilting the hybrid library increasingly to an electronic environment. Standardization, co-ordination of efforts and greater collaboration are the keys to a more problem-free programme of resource digitization, and Dawson discusses all of these based on his own experience in the UK.

Part 6

The final three chapters, forming Part 6, look at aspects of Standards and Technology in the merging digital environment. Perhaps most significant is the explosion of metadata, addressed by Dan Dorner of Victoria University of Wellington in Chapter 14 – 'Making sense of metadata: reading the words on a spinning top'. This chapter provides a synthesis of some of the key issues and developments related to metadata over the past four years, especially with reference to metadata for resource description and discovery. The aim of his chapter is to help information professionals become better informed about metadata and its roles, about the variety of metadata standards being developed and implemented, and their potential for enhancing access to digital collections. To achieve this synthesis is no easy matter, for the literature on, and the use of, metadata is expanding at a mind-boggling rate. In little more than a decade it has come from nowhere to occupy a major place in the discussions of information managers in all sectors, so it is appropriate to have such a well-written chapter to guide professionals through the metadata maze.

In Chapter 15 Christopher Brown-Syed writes on 'Beyond today's search engines'. His basic premise is that any sort of hierarchical organization cannot be applied to the web because of its inherently chaotic nature. Various suggestions have been offered to improve the situation, but in his view none of these is really viable or effective. On the horizon, however, are possible solutions which hold greater potential, including improved search engines, recourse to human experts, and the application of artificial intelligence. Brown-Syed also speculates that ant colonies, neural nets, faster computers and regional distributed web indexes all hold great promise for a less chaotic future in web-based information provision. And further over the horizon may lie effective multilingual support, adequate translation and speech recognition systems – all able to improve the technical functioning of the web for library and information access purposes. Brown-Syed's speculation makes one wish for the future to arrive even faster than it will!

In the final chapter Shadrack Katuu of the University of Botswana addresses a problematic, contentious and increasingly important issue – system security in an electronic environment. Ready access to information has long been the goal of the information profession, yet in an electronic environment we need to be equally concerned about the security of our resources and security of access – something that seems to contradict the principle of access. Katuu submits that we have no alternative but to find a viable compromise between open access and complete security, for without safeguarding our resources (both the hardware and the software) we run the risk of serious damage to our services. As he concludes, 'the risks involved in providing

access in an increasingly networked environment are both real and
potentially crippling if there are no safeguards for security'. His chapter suggests means of ensuring security without reducing the quality
of services.

The production team

As always, I owe a profound debt of gratitude to my highly professional Editorial Board for suggesting topics and authors, and for vet-
ting completed chapters. I also wish to thank, as usual, the team at
Facet Publishing, especially Rebecca and Helen, Lin, Alison and
Garry for their careful work in seeing this third volume through production. And, as always, Jackie Bell in Wellington has continued to
prop up my often over-taxed personal neural network. Finally, as most
of the volume was completed and edited while I was on sabbatical
leave in Beirut, I wish to thank my dear friends and 'support group'
in that marvellous city: Aida, Dolla, Gladys, Hilda, Maud and
Rudaynah. Good companionship and laughter always compensate for
hours at the computer screen – 'shukran, habibi'!

G. E. Gorman
Victoria University of Wellington

Part 1

IN PRAISE OF THE DIGITAL REVOLUTION?

1

The spectrum of digital objects in the library and beyond

Marilyn Deegan

Introduction

Libraries of the world exist to hold and preserve the human documentary heritage in whatever form that heritage might be represented and upon whatever medium it might be inscribed, and to make it available to a wide range of users. In the past these users have visited the library in order to interact with documentary witnesses in physical form, though there have been increasing efforts over the years to deliver information beyond the confines of individual libraries through interlending and other forms of document delivery. As Rebecca Lenzini states, 'It doesn't matter if you have it, it only matters if you can get it' (Lenzini, 1997, p.46). For users, the format of the documentary materials being accessed is generally irrelevant and only becomes an issue if access is impeded. What they need is convenience: the closest copy in the format most readily delivered, and this is increasingly being achieved by offering digital objects to library users. Most of the materials that libraries deliver tend to be available in multiple copies, but some items, especially in major research libraries, are unique, so user convenience has not been an issue; those wishing to view unique objects must visit the location and utilize them there – with all the problems for conservation that this entails. Now that many institutions are embarking upon programmes of retrospective digitization of rare and unique materials, the object – or a very good simulacrum of it – can be brought to the user rather than the user having to be brought to the object.

Libraries have long been early adopters of new technologies if these can help them achieve the mission of the organization; the use of administrative systems for catalogue record creation is, for instance, an early example of the use of computerized processes in libraries. Alongside this new capability to share catalogues, there started co-operative efforts to share the cataloguing load and to benefit from aggregating cataloguing efforts across multiple libraries. Such developments in co-operative efforts for library processes were enabled by the essential 'share-ability' of the digital medium, and have extended greatly in the development of digital libraries. Many of the major programmes of digital library development are collaborative ones, with much of this collaboration being on a worldwide basis. This is bringing many benefits to the library world, but there are also some concerns for those who may be excluded from this brave new world of digital activity because of various demographic and developmental reasons. There exists, some say, a digital divide between young and old, rich and poor, male and female, and developed and developing nations. This chapter examines the background to digital library developments, looking at the nature of digital objects, the changing nature of libraries as organizations, and some of the international issues in collaboration and access which can both help and hinder the wide uptake of digital library developments.

Digital developments

The formats of documents and other materials to be found in libraries are many, from printed books, journals and other publications, to manuscripts written on paper, vellum, papyrus, birch bark, wood and many other substrates. There is also a whole range of surrogate formats, including photographs on glass plates, negative film, photographic paper, microfilm, microfiche, as well as an increasing range of audio-visual materials. Librarians are used to dealing with a hybrid world of documentary formats, and to providing the technologies through which these can be accessed. Libraries have always been hybrid organizations (though the term 'hybrid library' is coming to be used to describe the conjunction between digital and analogue objects), and this 'hybridicity' has not really changed the nature of libraries in the past. Increasingly, though, it is believed that the rapid advance of the digital *does* change the nature of libraries, and that digital libraries are in some way fundamentally different from traditional libraries – though defining what constitutes a traditional library is extremely difficult to do. It is equally difficult to define a digital library, though many attempts have been made. A set of defining principles that seem unarguable is offered in Deegan and Tanner (2002, p.22):

- A digital library is a managed collection of digital objects.
- The digital objects are created or collected according to principles of collection development.
- The digital objects are made available in a cohesive manner, supported by services necessary to allow users to retrieve and exploit the resources just as they would any other library materials.
- The digital objects are treated as long-term stable resources and appropriate processes are applied to them to ensure their quality and survivability.

There are those who feel, however, that using the term 'digital' in 'digital library' is a redundancy, and Braude, for instance, points out the difference between the 'product that we manage in libraries, information, and the familiar container for that product, the codex book' (Braude, 1999, p.85). These containers have influenced library architecture, but they do not themselves define what a library is. Braude suggests that, as 'we did not bother to qualify our libraries by calling them clay libraries or papyrus roll libraries, why now do we have to call them digital libraries?' (Braude, 1999, p.86). These comments are valid and, in terms of the intellectual management and understanding of data, need to be taken seriously, but there *is* a distinction to be made between traditional libraries and digital libraries (or digital collections within libraries). Physical containers for information are capable of direct access and can be managed physically – they are stored in environments best suited to their particular needs and delivered physically to the users for access. Digital data is made up of electronic signals that rely on an interpreting machine before there can be any human interaction with it. Without this machine it is impossible to access; it is ephemeral, fragile, can be detached from its 'container' with (usually) minimal effort (which works inscribed in codices cannot), can be replicated indefinitely almost to infinity, can be altered without trace.

While the differences between analogue and digital data may be of degree more than of substance, they are sufficiently large to require different approaches. Libraries have to manage digital data alongside all the other kinds of information objects they manage, and they also have to manage the machines – a new departure for them. This has huge implications for the economics of libraries, it has implications for the profession of librarianship, and it has consequences for the relationship between library and user. The virtual nature of the digital medium is also a key issue for libraries: no longer is a library bounded by its physical walls; its data arrives in digital formats from suppliers all over the world; and it can itself supply digital copies of its own materials where and to whom it wishes, within the limits of the legal issues which now surround the digital world, and the economic needs of and constraints upon the library.

The spectrum of digital objects

While all digital objects are made up of a series of electronic impulses (bits or BInary DigiTs) which have only two states ('on' or 'off', represented by '1' or '0'), these impulses can be combined in many different ways to represent the whole spectrum of library objects. Most libraries will be concerned largely with electronic text or digital images, though increasingly they will need to handle digital sound and digital video – indeed there are already a number of specialist libraries faced with these formats. The greatest volume of digital material that libraries access is almost certainly brought in from outside: purchased or licensed from publishers or third-party suppliers, or delivered free via the world wide web. Most of this data is now delivered and accessed through networks, though some is still supplied on CD-ROM.

The CD-ROM has proved to be a relatively transient medium for libraries, and with its decline comes an interesting problem in the definition of digital objects: where are the boundaries? In the inter-net-worked world, what *is* a digital object? With CD-ROMs (or even floppy disks before them), an object was purchased or licensed which could be catalogued and (in the early days) handed over a counter to a user, who then accessed it on a machine inside the library, or even borrowed it for use outside. Then CD-ROMs were networked, and now information is delivered direct, without being supplied on a physical object. The largest growth area in the delivery of library data is currently online journals, and this is a particularly interesting domain within which to ask the question posed above: what *is* a digital object? Is a journal article a digital object? Is it the whole journal issue? The title? The service offered by one supplier? The answer is, of course, that in different contexts these all can be considered as individual digital objects – it depends on one's point of view, and it might also depend on what one is paying for. The perspective of a user paying to be able to download one article is likely to be different from that of an organization paying one subscription to a service that offers hundreds of journals.

Delivering digital objects

Digital objects come into libraries in a variety of formats, all of which need to be delivered to the library user in a way that is apparently seamless. One of the main problems with CD-ROMs in the past has been the number of interfaces and interaction mechanisms which the librarian and user had to learn. Publishers became software rather than information suppliers and perceived market advantage to be gained from innovative programming rather than unique or useful content. This created a Tower of Babel of nightmare proportions for the delivery and preservation of library materials, particularly if that content was no longer supplied in printed form. The move to online supply, and the realization that it is content that has value for suppliers, has eased this problem somewhat, as has the insistence in the library world upon the development of underlying standards. However, as the old joke goes, 'the good thing about standards is that there are so many of them to choose from', and there has been only partial success in the application of standards in the fast-moving world of digital data creation and use.

Some attempts have been made to understand the problems in applying standardization in this area (see Gill and Miller, 2002), and there are a number of international bodies addressing the issues. But innovation and standardization are to some degree antithetical, and there is still a great deal of welcome innovation in the digital library world. The recent and rapid rise in popularity of the e-book, for instance, is a case in point. There are several different formats of e-books and e-book readers, and they are incompatible because there is no pre-existing standard for such a new format. Market forces will probably ensure that one standard will prevail over the others eventually (experience attests to the fact that it will not necessarily be the best one), but libraries will be faced with having to deliver a number of parallel formats for some time. A promising development is in the area of interoperability, where work is being done to create crosswalks between standards – mappings which allow data created using one standard to be interrogated alongside data created with another, though there is still much to do in this area (see Miller, 2000).

The spectrum of digital objects that the library needs to deliver currently is broad, and, as noted above, there are always new innovations that further increase the range, at the same time as earlier developments become more mature and standardized. This means that libraries need to accept that the situation will be complex and shifting for some time and that future stability is a chimera. The present range of digital objects comes from myriad sources, including:

- the library's own holdings that have been digitized
- purchased datasets on CD-ROM
- purchased datasets that are online
- electronic publications that have a paper equivalent
- electronic publications that have no paper equivalent
- electronic reference works, which increasingly have no paper equivalent
- e-books.

These have to be delivered by the library alongside the non-digital holdings, and also integrated with both the digital holdings of other libraries that might be of value to local users, and with the internet and its myriad resources. All of the above are in many different formats, with different metadata standards applied to their description. They have to be catalogued, in itself a challenge and, depending on the function and status of the institution, preserved for long-term access and sustainability. A number of national libraries are grappling with the issues of the preservation of published electronic materials offered by publishers on voluntary or legal deposit (the British Library and the Royal Library in Holland are two such), and others are investigating the preservation of more ephemeral data such as websites as well (the Library of Congress in the USA, the National Library of Australia, and libraries in Sweden, Finland and Denmark).

Formats of digital objects

Digital objects deriving from the sources listed above come in a variety of formats, representing a wide range of analogue originals.

Increasingly they derive from digital originals, the so-called 'born digital' materials which have no analogue or print antecedent. The following are some of the major formats that currently need to be dealt with:

- electronic text formats, including electronic full text with mark-up, electronic text in portable document formats (PDF), electronic text as page images
- digital images
- photographic collections
- electronic books (e-books).

Electronic text formats

Electronic text has been around as long as computers have – more than half a century. Early processors manipulated alphanumeric symbols, not images, so text was relatively straightforward to process, though very difficult to enter into the machine, early data entry being via punch cards or tape. A number of projects emerged which re-keyed printed originals in order that some manipulations could be carried out on them for linguistic or stylistic purposes. Works that were converted at this early stage included the writings of Thomas Aquinas, the entire corpus of Old English literature and the corpus of classical Greek literature. Note that not all these are in the standard ASCII character set. A special encoding system was devised for classical Greek so that the full range of characters, breathings and accents could be represented by the limited symbols of ASCII and interchanged in standard form. These early electronic texts were produced as tools for the generation of another analogue derivative, such as a new printed edition, or an analytic book or article, but the value of the text, and more particularly its reusability for other purposes, meant that what was originally part of a process became itself a product. During the 1970s, computerized typesetting was employed to produce print originals by publishers, at the same time as word-processing was being developed for business and personal use. The worlds of publishing and commerce were soon overwhelmed by electronic text, to the extent that pundits confidently predicted the death of the printed book. It is in the area of electronic text that the most successful standards have emerged. Early textual formats were developed for local processing and production needs, meaning that word-processing standards were different from typesetting standards (and indeed even typesetting standards differed between publishers), and projects capturing text for particular manipulations would devise schemes of their own. Some systems were used for presentation, others for analysis, and this fluidity of formats meant that the great benefits of potential reusability and interchange of information were not yet able to be exploited fully.

In the 1970s a system was developed by researchers at IBM to create textual formats that were generalized; no longer was mark-up embedded in text to say what should happen to the text or how it should look, rather mark-up defined the ontology of textual objects. That is, it stated what these objects were, a 'heading of level one' say, rather than 'a heading, Times Roman, Bold, 24 pt'. The publishing industry rapidly adopted this family of standards and in the 1980s SGML (Standard Generalized Mark-up Language) was developed and used widely throughout the publishing industry. It is this adoption of standards that has fired the exponential growth in electronic publishing, with hundreds of thousands of major journals and other publications now widely available in electronic form. SGML has many benefits, but it is highly complex, so a drastically pruned version of it, HTML (Hypertext Mark-up Language), was developed as a mark-up scheme for the exchange of documents for the world wide web in its early stages. this simple standard which was easy and quick to learn has given rise to the many billions of websites now available, but has been cracking under the strain of the uses to which it has been put – uses never planned by its creators. Therefore a new standard has been produced that is also based on SGML, simpler than its parent but richer than HTML: XML (eXtensible Mark-up Language) is now the accepted format for electronic documents and other digital objects.

Electronic text marked up in SGML, HTML or XML is searchable and flexible, but it has certain disadvantages for publishers. It is so amenable to manipulation and easy to exchange that there is a loss of stability in information delivery, and there are problems of authenticity. An electronic text can be altered silently, with no trace of where the alteration occurs or of who did the altering. Without meticulous attention to recording all changes, version control is problematic, and citation becomes unreliable. When printing an HTML file, for instance, page breaks are a matter of local printer settings rather than some fixed, edition-specific feature.

Now popular with publishers is the portable document format or PDF. This preserves the integrity of the page across systems and gives excellent results for printing; furthermore, PDFs cannot easily be altered by users, so version control and authenticity are less of a problem. PDF originated as a page description language, and as such it serves publishers well, giving them full control over how documents will look. However, there are limitations in the searching and manipulation of PDF text, though recent versions have improved tools for search and retrieval. The most serious problem with the PDF format is that it is proprietary and therefore not an open standard, giving the suppliers of the format, the Adobe company, too much control. However, there is now so much material available in PDF that it is unlikely that the company will either collapse or substantially alter the format to the detriment of its users: that would not be a good business strategem, and so PDF has become a de facto standard in the online world, particularly for presentation purposes.

When creating digital versions of materials that are originally in printed form, one widely used method of capture and delivery is the digitization of printed pages as images, and then the use of optical character recognition (OCR) tools for the creation of searchable indexes. Depending on the quality of the printed original, OCR accuracy can be relatively high and can therefore give good retrieval with search tools. Even where texts are older or in poor condition, there are now software tools available, produced using artificial intelligence

techniques, that can give acceptable indexing from relatively compromised originals. In the world of academic libraries, these techniques were pioneered by institutions such as Yale and Cornell with
projects to conserve and reformat brittle books through microfilming
and digital imaging, and by the Making of America Project at Cornell
and Michigan (see Chapman, Conway and Kenney, 1999,
http://moa.cit.cornell.edu and http://moa.umdl.umich.edu/). This
method has also been used in the digitization of back runs of journals by projects like JSTOR and MUSE, where fidelity to the printed
originals is vital (www.jstor.org/; http://muse.jhu.edu/). Great
strides, too, have been made in the delivery of acceptable digital fac-
similes of even such difficult formats as historic newspapers, with
sometimes astonishing results for searchability, as for instance
achieved by the British Library Newspaper Pilot which uses Olive
Software's ActivePaper Archive (www.uk.olivesoftware.com).

Digital images

Most libraries are rich in visual data in their documentary collections,
and many have substantial holdings of photographic materials. A
great deal of the earlier and rare materials in library holdings is of as
much interest for its visual appearance and its artefactual status as it
is for its content. Advanced digital imaging techniques can now capture images of originals that satisfy even the most intensive scholarly
research, and provision of high-quality surrogates allows access to
fragile originals to be restricted in order to prevent further deterio-
ration. Libraries that offer reprographic services are finding that digital capture has many advantages over photographic reproduction,
and high-quality photo-realistic prints can be made available from the
digital files alongside a whole range of digital delivery routes. Indeed,
many libraries and other cultural institutions offer images of medium
quality free for non-profit uses such as education. The Oxford Digital
Library and the Victoria and Albert Museum in the UK and the
Library of Congress in the USA offer such free usage.

Digital images in libraries have a huge range of benefits for both libraries and users. Manuscript images, for example, can be manipulated to reveal hitherto obscure or damaged features; the Electronic Beowulf project carried out by the British Library and the University of Kentucky has revealed readings that were formerly hidden under earlier repair work to the manuscript (www.bl.uk/collections/treasures/beowulf.html). Similarly, the Digital Image Archive of Medieval Music project (DIAMM) has enhanced images of medieval music manuscripts to reveal lost musical works hidden underneath palimpsest texts (www.diamm.ac.uk). Digital imaging of library collections can also allow related items to be brought together on one screen for comparison, even though they may be geographically dispersed. It is now possible, for example, to access online or on CD-ROM seven copies of the remaining 48 copies or fragments of copies of the Gutenberg Bible. These seven copies are located in Keio University, Tokyo; Göttingen; Mainz (two copies); the British Library (two copies); and Cambridge (see www.humi.keio.ac.jp/ and www.sub.uni-goettingen.de/gdz/). The availability of these for instant online comparison is a huge benefit for the history of printing. No two copies of the work are exactly the same, as illuminations and rubrication were added by hand, and the printing is not entirely consistent from volume to volume.

Photographic collections

Many libraries and other cultural institutions hold photographic collections large and small, and these are increasingly at risk because of inherent instabilities in the photographic process; even relatively recent materials like colour prints of the last ten years or so are fading and losing definition. Many of the holdings are in urgent need of conservation and reformatting, and some of the materials are actually dangerously unstable, such as nitrate film stock, which can ignite or even explode without warning. A survey of photographic materials in Europe carried out by the European Commission on Preservation and Access (www.knaw.nl/ecpa/) revealed that the 140 responding institutions hold some 120 million photographs, half of which are over 50 years old. This is an astonishing number and gives some sense of the scale of the world's photographic stock, all of which is at risk. Around 80% of the survey respondents had already begun digitization of their photographic holdings or were planning to digitize in the future, with protection of vulnerable originals being a crucial impetus.

Digital surrogates are excellent substitutes for photographic originals and can provide almost all the content information that the original can yield, though of course artefactual information cannot be conveyed other than in the metadata. The Pictures Catalogue at the National Library of Australia has some 30,000 digital images of Australiana, many of these from photographic originals (from a total collection of 550,000 photographs which the Library holds). These are added to regularly (www.nla.gov.au/catalogue/pictures/). The Library of Virginia Digital Collections have more than 10,000 digital images of photographs on a variety of topics of interest to the citizens of Virginia (www.lva.lib.va.us/dlp/index.htm). In the UK the JISC Image Digitization Initiative (JIDI) has digitized around 50,000 images, many of them from photographic collections. These are described at (www.ilrt.bris.ac.uk/jidi/).

Electronic books

An exciting but problematic new development in the digital library world is the electronic book or e-book. This differs from an electronic text in that it has formatting which makes it more like a printed book to use, and e-books purchased from commercial suppliers often have internal coding which restricts their use only to the registered machine for which they were originally supplied. The paradox here is that digital data which is inherently fluid, interchangeable and infinitely replicable can, for legal and economic reasons, be controlled more rigidly than the printed book, which can be loaned privately or publicly and sold on whenever the original purchaser wishes. One company, RosettaBooks (www.rosettabooks.com/), even offers a time-limited product, a self-destructing e-book, which is electronically available at low cost for up to ten hours before the content disappears and has to be renewed (Bowman, 2001).

E-books can be read on PDAs or on full-size computers, but many have been designed for use with specialist readers such as the RCA Gemstar or the Everybook. These are surprisingly ergonomic and pleasant to use, combining comfort for hand and eye with some limited added functionality: searching, marking, annotation, linking to dictionaries and encyclopedias. The main disadvantage of e-book readers is the price – the cheapest retail at around US$300, and books purchased for use on them are around the same price as the hardback versions. There is a good choice of titles available, however, and a number of libraries have experimented with the purchase of e-book readers and their supply to readers, with a good deal of success (Cox and Ormes, 2001).

Another major development in the supply of e-books is the rise of such companies as Questia and netLibrary which supply e-books through the web as a service for a regular fee rather than as individual titles. These are mostly aimed at a student market. Questia targets individuals directly, while netLibrary works with libraries and has recently been purchased by OCLC, which should strengthen its relationship with the libraries community.

Access for all?
The demographics of digital data

One of the great shibboleths of our time is that the digital medium is essentially democratizing. But less than 6% of the world's population is connected to the internet, and in some parts of the world, Africa for instance, the vast majority of the population – around 80% according to Taylor (2001, p.35) – has never even made a telephone

call. In the poorer parts of the world connectivity is not just dispro-
portionately expensive, it is expensive even by developed Western
standards. Poor general infrastructure (electricity, cabling, telephone
communications, water) means that establishing networking is both
difficult and a low priority for those for whom obtaining even the
necessities for survival is a struggle. It is a sad paradox that a medium
lauded as so inherently shareable and that has the potential to tran-
scend boundaries so easily is not accessible to most of the world.

Even within the developed world, access to digital resources is
demographically skewed towards the young, the male, the middle
class, and the anglophone. But the balance is slowly shifting as more
materials become available which appeal to an ever wider range of
users, and as internet cafés, public libraries and other public spaces
offer free or low-cost access, and some service providers give free e-
mail accounts. Growing numbers of women and the elderly (known
as 'silver surfers') are now using the internet, and there are some
promising inventions, including new methods of wireless communication like the Bluetooth protocol, that should offer improvements
for the future of connectivity in the developing world; but cost is
always likely to be an issue. The fear is that the information revolutions will widen the gap between the developed and developing
worlds, as the poorer nations fall further behind in economic growth
and education.

What is needed is a will to innovate and change, and to imagine
creative solutions to the problem. Developing nations have a paucity
of educational materials and resources, and digital libraries could be
revolutionary and truly democratizing with the right social, commercial, political and educational partnerships. An exciting new initiative
has just been announced by the World Health Organization: publishers of '1,000 leading medical and scientific journals will offer electronic subscriptions at "free or deeply-reduced" rates to institutions
in the developing world' (www.who.int/inf-pr-2001/en/pr2001-32.
html). The delivery of these journals is mutually beneficial in that it
costs the publishers virtually nothing to make them available in digi-

tal format for the three years initially agreed, but they gain significant market knowledge and customer loyalty. The developing nations gain access to an information resource they were very unlikely to have been able to afford previously.

Another interesting possibility is being promoted by the New Zealand Digital Library with the development of the Greenstone digital library software. This is public domain software made available under the GNU General Public Licence, and it is supported by UNESCO. The GNU General Public Licence is available online and is designed to 'guarantee the freedom to share and change free software – to make sure the software is free for all its users', and applies to 'any program whose authors commit to using it' (Free Software Foundation, 1991). It has been used for delivery at extremely low cost of important materials on CD-ROM and the internet, including the Online Burma Library (www.burmalibrary.org/) and the Humanity Digital Library, 'a large collection of practical information aimed at helping reduce poverty, increasing human potential, and providing a practical and useful education for all' (www.nzdl.org) (Witten, Bainbridge and Boddie, 2001). The Forced Migration Online project at the Refugee Studies Centre, University of Oxford, is producing digital data on all issues to do with human displacement that will be made available as widely as possible in the developing world (www.forcedmigration.org). The Digital Imaging Project of South Africa (DISA) is developing tools for capacity building in digital data and digital methods throughout South Africa, with support from the Andrew W. Mellon Foundation. This project aims to make materials relating to South Africa available to scholars around the world, as well as giving librarians and archivists in South Africa some experience in digital imaging (Peters and Pickover, 2001). The African Digital Library being developed by a number of universities in Africa, with Michigan State University in the USA and funding from the US National Science Foundation, is developing a multimedia digital library of West African sources in multiple languages for the benefit of scholars within and outside Africa (http://africandl.org/index.html).

The print-disabled

One of the great benefits that digitization can bring to libraries is the possibility to extend their reach beyond the traditional user base. Traditionally, many libraries have had little to offer those who cannot read or hold printed materials – the print-disabled – but new developments are creating new possibilities for equalizing access to printed materials and enabling new educational and recreational activities. There is now a set of principles known as 'design-for-all' which suggest:

> that library IT systems and interfaces are designed in a way that can be easily read by all users of the library, be they physically visiting the library itself or accessing it remotely and regardless of any disability or access preference they may have. The RNIB describes design for all in relation to websites as 'a single version of the website which is accessible to everyone, however they access the Internet'.
>
> (Brophy and Craven, 2001)

A useful tool for testing whether design-for-all principles have been followed is Bobby, described as 'a Web-based tool that analyses Web pages for their accessibility to people with disabilities' (www.cast.org/Bobby/Bobby311.cfm). This also suggests improvements that might be made, such as accompanying visual materials with a sound track, and the suggestions help web authors optimize web pages for use with special browsers and text-to-speech converters. Another exciting development for the print-disabled is DAISY, the Digital Audio-based Information System, which is used to create digital talking books with more sophisticated access mechanisms for complex documents than is possible with linear recording technologies (www.daisy.org).

Realizing the new opportunities that the digital medium offers for the extension of library services to the print-disabled, a group of European libraries and organizations for the blind formed a consortium to produce customized workstations and training materials for the use of blind or print-disabled students. The Accelerate Project was funded by the European Union and ran for two years from 1999 (www.lib.uom.gr/accelerate/). While to the rest of the world the development of GUI interfaces was a great step forward in navigation of complex information spaces, to the print-disabled this was a disaster, reliant as they had become on text-to-speech processing technology which works admirably in a text-only environment. The Accelerate Project (with partners in Greece, Cyprus, Austria and Holland) developed adapted workstations for print-disabled students, plus training manuals for librarians who need to understand the special needs, and user manuals to train students in the navigation of the tools of a modern academic library – the library catalogue, databases (online and CD-ROM), the internet – and the adaptive equipment which provides the access to the tools.

Conclusion

There are many different kinds of organization that use the term 'digital library' to describe what they are developing, and digital libraries can use any number of organizing principles even while following the tenets offered at the beginning of this chapter. A digital library can be generated from the holdings of one particular institution, such as the Oxford Digital Library (www.odl.ox.ac.uk) and the digital library of the Czech National Library (http://digit.nkp.cz/). It can be organized around a theme like the Online Burma Library mentioned above or the Tibetan and Himalayan Digital Library (http://iris.lib. virginia.edu/tibet/frameset.html). Particularly exciting is the possibility for collaborative projects and programmes to develop materials and methods throughout the world, and a number of funding bodies have had the vision to support these. They include the National Science Foundation in the USA, the European Union, the Joint Information Systems Committee in the UK, and the Andrew W. Mellon Foundation which funds developments across a broad geographic area. If we can solve the access problems outlined above, and many are trying to, then the sharing of materials, expertise and scholarship that these programmes offer is truly exciting.

The digital world is moving fast, and there are many information services competing for the territory traditionally occupied by libraries and librarians. However, this should not be a cause for alarm but for new opportunities – never have librarians been needed more for helping users find and evaluate information sources. As Rennie (1997, p.6) points out, 'at some point the Internet has to stop looking like the world's largest rummage sale. For taming this particular frontier the right people are librarians, not cowboys. The Internet is made of information and nobody knows more about how to order information than librarians.' It is important to remember, too, that it is *information* that is important, not *format*. Digital supply can be fast and convenient, but in the end it is quality of content that matters, and quality content is what libraries do well. Libraries have embraced the digital world with fervour, and some of the most exciting developments in computing and information management are happening in the library world, as the rest of this volume demonstrates.

References

African Online Digital Library
   http://africandl.org/index.html
   (accessed 21 April 2002)
Bobby worldwide tool for web authors
   www.cast.org/Bobby/Bobby311.cfm
   (accessed 21 April 2002)
Bowman, L. M. (2001) Disappearing ink: e-book self-destructs, *cnetNews.com*, (8 August), available at
   http://news.com.com/2100-1023-271303.html?legacy=cnet
   (accessed 21 April 2002)
Braude, R. M. (1999) Virtual or actual: the term library is enough, *Bulletin of the Medical Librarians Association*, **87** (1), 85–7.

British Library. Electronic Beowulf Project
www.bl.uk/collections/treasures/beowulf.html
(accessed 21 April 2002)

British Library. Newspaper Pilot
www.uk.olivesoftware.com
(accessed 21 April 2002)

Brophy, P. and Craven, J. (2001) Accessible library web sites: design
for all. In *Library services for visually impaired people: a manual of best
practice*, London: National Library for the Blind, available at
www.nlbuk.org/bpm/contents.html
(accessed 21 April 2002)

Chapman, S., Conway, P. and Kenney, A. R. (1999) Digital imaging
and preservation microfilm: the future of the hybrid approach for
the preservation of brittle books, *RLG DigiNews*, available at
www.thames.rlg.org/preserv/diginews/diginews3-1.html
(accessed 21 April 2002)

Cox, A. and Ormes, S. (2001) *E-books*, Library and Information
Briefings, 96, London: South Bank University, available at
http://litc.sbu.ac.uk/publications/libs/libs96.pdf
(accessed 21 April 2002)

Czech National Library. Digitization Programme
http://digit.nkp.cz
(accessed 21 April 2002)

DAISY, the Digital Audio-Based Information System
www.daisy.org
(accessed 21 April 2002)

Deegan, M. and Tanner, S. (2002) *Digital futures: strategies for the information age*, London: Library Association Publishing.

European Commission on Preservation and Access
www.knaw.nl/ecpa/
(accessed 21 April 2002)

Forced Migration Online
www.forcedmigration.org
(accessed 21 April 2002)

Free Software Foundation (1991) GNU General Public Licence, available at
www.gnu.org/copyleft/gpl.html
(accessed 3 July 2002)

Gill, T. and Miller, P. (2002) Re-inventing the wheel? Standards, interoperability and digital cultural content, *D-Lib Magazine*, **8** (1), available at
http://mirrored.ukoln.ac.uk/lis-journals/dlib/dlib/dlib/january02/gill/01gill.html
(accessed 21 April 2002)

Johns Hopkins University. Project Muse
muse.jhu.edu/
(accessed 21 April 2002)

Joint Information Systems Committee. Image Digitisation Initiative
www.ilrt.bris.ac.uk/jidi/
(accessed 21 April 2002)

JSTOR, journal storage: the scholarly journal archive
www.jstor.org)
(accessed 21 April 2002)

Keio University. Gutenberg Project. Humanities Media Interface Project (HUMI)
www.humi.keio.ac.jp
(accessed 21 April 2002)

Lenzini, R. T. (1997) Having our cake and eating it too: combining aggregated and distributed resources. In Lee, S. H. (ed.) *Economics of digital information: collection, storage and delivery*, New York: Haworth Press, 39–48.

Library of Virginia Digital Collections
www.lva.lib.va.us/dlp/index.htm
(accessed 21 April 2002)

Miller, P. (2000) Interoperability: what is it and why should I want it?, *Ariadne*, **24**, available at
www.ariadne.ac.uk/issue24/interoperability/intro.html
(accessed 21 April 2002)

National Library of Australia. Pictures catalogue
   www.nla.gov.au/catalogue/pictures/
   (accessed 21 April 2002)
New Zealand Digital Library
   www.nzdl.org
   (accessed 21 April 2002)
Online Burma Library
   www.burmalibrary.org/
   (accessed 21 April 2002)
Peters, D. and Pickover, M. (2001) DISA: insights of an African model
   for digital library development. *D-Lib Magazine*, **7** (11), available at
   www.dlib.org/dlib/november01/peters/11peters.html
   (accessed 21 April 2002)
Rennie, J. (1997) Civilizing the internet, *Scientific American*, (March), 6.
RosettaBooks
   www.rosettabooks.com/
   (accessed 21 April 2002)
Taylor, D. J. (2001) Curse of the teenage cybergeeks: review of
   Michael Lewis, *The future just happened, Sunday Times Culture*, (15
   July), 34–5.
Tibetan and Himalayan Digital Library
   http://iris.lib.virginia.edu/tibet/frameset.html
   (accessed 21 April 2002)
University of Göttingen. Gutenberg Digital Library
   www.sub.uni-goettingen.de/gdz/
   (accessed 21 April 2002)
*University of London. Royal Holloway College and University of Oxford.
   Digital Image Archive of Medieval Music*
   www.diamm.ac.uk
   (accessed 21 April 2002)
University of Macedonia. Accelerate Project
   www.lib.uom.gr/accelerate/
   (accessed 21 April 2002)

University of Michigan. Making of America
   http://moa.cit.cornell.edu/ and http://moa.umdl.umich.edu/
   (accessed 21 April 2002)
University of Oxford. Oxford Digital Library
   www.odl.ox.ac.uk
   (accessed 21 April 2002)
Witten, I. H., Bainbridge, D. and Boddie, S. J. (2001) Greenstone:
   open-source digital library software, *D-Lib Magazine*, **7** (10), available at
   www.dlib.org/dlib/october01/witten/10witten.html
   (accessed 21 April 2002)
World Health Organization (2001) Press release on online journals
   deal, available at
   www.who.int/inf-pr-2001/en/pr2001-32.html
   (accessed 21 April 2002)

2

Digital versus print issues

Lorna Peterson

Introduction

Digital versus print issues are often presented in a highly charged, emotional language, using a rhetoric of progress and optimism or fear and loathing. This chapter provides an overview of the subject, addressing the opposing viewpoints, the cultural-historical context and the business influences on the arguments in the digital versus print debate. It addresses the individual and institutional issues regarding digital versus print formats. Recognizing that the foundation of media librarianship embraces the notion of content, not form, in the selection, acquisition, management and use of information, the position in this chapter is that the debate is spurious. The debate is characterized as spurious because underlying, but rarely articulated, in the issue of digital versus print are the concerns of status, cost and values. Subtle, but apparent, these concerns manifest themselves as:

- *status* – appearing progressive, cutting edge, leading, risk taking and therefore superior; no caution about technology and its stage of development or feasibility
- *cost* – expensive and constantly needing upgrades, technology purchases in libraries are often made at great sacrifice, particularly to the print collections
- *values* – the why and worth of pursuing a particular direction.

The assumed medium under discussion is written communication, whether monographic or serial. It is either fiction or non-fiction, research or scholarship, news reporting or general interest, and is targeted at a specific group or combination of audiences such as children, adolescents, adults, experts, amateurs, etc. Unless otherwise noted, the term 'book' and its digital counterpart, 'e-book', will be used as shorthand for these types of publications. The discussion of reading, preserving, accessing, collecting, purchasing and using digital or print information assumes that the form and purpose of newspapers, scholarly journals, magazines, textbooks, scripture, devotional literature, manuals, reference books, etc. can be addressed in a general way to include the common issues under contention in the digital versus print debate.

Cultural and commercial conflict

Published, written communication, especially in the form of a monograph that is fiction or non-fiction, carries enormous cultural cachet. Books and the book life are often romanticized. Books are often not viewed as part of the business world or as a commodity (Greco, 1997). An entire literature is devoted to the meaning of books to individual lives (Schwartz, 1996; Toth, 1991), which elevates the book's role in society. As a medium for intellectual and creative outlets, the book conjures up feelings of sacredness because its contents often deal with truth, beauty, imagination, faith, ideas and knowledge, providing inspiration, information and entertainment. But books are products, and publishing is a commercial enterprise driven by profits and competition; therefore, product development, marketing, advertising and other forms of business manipulation are as influential in the shaping of how consumers desire and purchase written published communication as in how consumers desire and purchase soap, automobiles or other consumer goods.

This is an important distinction to make, because the selling of new products often relies upon creating a need, overstating the product's capability and taking advantage of human foibles such as insecurity, anxiety, greed and materialism. New product development also springs from invention and innovation, and the human qualities of curiosity, intelligence and imagination help engineer our new products and provide progress, comfort and convenience; in this sense, product development or new technology should be praised and generally not feared.

The development and sale of digital reading matter must be considered in this context to understand fully the promotion of digital forms as superior and the death of the traditional book as imminent. Unfortunately, the hyper-selling commercial context of published communication creates a false dichotomy of digital against print rather than fostering an ethos of co-existence and opportunities for choice. Market values often triumph over cultural and intellectual values not because they are right, but because the cultural and intellectual values are dismissed, silenced and left with no compromise. The digital versus print argument frequently touts digital forms as superior, necessary, progressive and therefore unavoidable. Those who favour print frequently cite tradition, aesthetics and cost. Rarely does either camp acknowledge market manipulation inherent in the discussion. Even less frequent is the acknowledgement that both can and should co-exist and usage should be according to consumer choice. For librarians and libraries the consequence of not truly having a choice erodes the opportunity of having the best of both. In order to be perceived as worthy of financial support, librarians often embrace technology and offer technological services to show the administrators and politicians who are responsible for the institution's fiscal health that they are on the cutting edge. But too often the use of technology in development predates its feasibility. Technology, always costly, may not deliver on its promises, and caution is warranted. It is this tension that causes the debate, but the format of material is easier to attack.

The historical context

Revolutions or evolutions?

Popular history has long recognized the movable-type printing press as a revolution in communication. The computer as a communication device is seen as having a similar revolutionary impact on society. But book history and print culture are questioning the popular notions of a printing press revolution. The effects, nature and significance of the printing press are under debate, being reinterpreted and revised, while acknowledgement is being given to the technological, economic, social, political, artistic and religious factors that intervene between writer, publisher and the printed text, and the consequences of that intervention. Those who speak of a computer revolution in communication would do well to visit the conflicting interpretations of the printing press revolution in order to understand the complexity of the issues and gain deeper respect for the concerns. The digital versus print debate is not a simple matter of tradition versus progress.

Are these technological advances (printing press or computer) revolutions or evolutions? Elizabeth Eisenstein (2002) studies the way the transmission of texts over history changes historical consciousness. Johns (2002) integrates book history with intellectual and cultural history to form a new historiography of print studies. The *American Historical Review* has presented a forum titled 'How revolutionary was the print revolution?'(Grafton, 2002), suggesting that there is much we do not understand regarding social and scientific transformations, regardless of the communication medium that facilitated them. The same is true for the so-called digital information revolution.

Miniaturization of information: microforms, videodiscs, DVDs and CD-ROMs

As paper and print became the standard form for written communication, their reproduction and reduction became major issues. The need to micro-reproduce was seen as a way to save precious shelving space, to preserve and protect essential information, and to process information. Microform, videodisc and CD-ROM technologies presented advantages and disadvantages to readers and libraries. Following World War 2, the use of microforms as information sources and storage devices was widely embraced by libraries. The arguments in favour of using microforms were: 1) they saved space, 2) they increased access and 3) they preserved fragile items (Benedon, 1978). Use of microforms in libraries resulted in a debate similar to the recent digital versus print argument – users did not like the form but were grateful for the access; equipment costs and maintenance were detriments; new staffing arrangements had to be made as well as circulation and borrowing policies developed. Nicholson Baker (2001) critically chronicles microfilming and microforms in libraries. Richard Cox (2000) counters Baker's work with a warning to librarians on the likelihood that, in this digital age, a backlash of intense concern by the general public for original books, archival materials and other documents will result.

Another non-paper and print form experimented with in libraries was videodisc. These laser discs of video clips, sound recording, maps, text, etc. are multimedia presentations of information. They were late 1970s inventions, which matured by the mid-1980s, and were once considered cutting-edge replacements for traditional books. Videodiscs, and the later DVD (Digital Video Disc or Digital Versatile Disc), have not replaced traditional books but have supplemented them.

CD-ROM technology entered libraries in the form of multimedia resources such as encyclopedias, information banks of numeric data such as census materials, bibliographic electronic retrieval databases and other information forms. As a non-print form, CD-ROM democratized online bibliographic searching. Online searching was once mediated by a librarian, with the search results paid for by the client. CD-ROM technology democratized online searching by eliminating the mediator and passing the cost on to the institution. CD-ROM technology also provided some of the earliest full-text access.

Databases such as SIRS, Inc. provided, in full-text form, carefully selected articles on social controversies from general interest and scholarly journals. The full text was helpful to smaller libraries that could not grow in space and maintain deep periodical collections. Full text on CD-ROM also helped overcome the problem of novice bibliographic users not finding materials because of binding, re-shelving, or non-subscription. In the debate about CD-ROM, as with the other formats, the equipment cost, maintenance, technical knowledge and perceived diminished importance of print were cited as negative factors. Saving space and increasing access were seen as the benefits.

The arguments: digital versus print

Addressing the positive and negative aspects of various ways to access and use information, including print and computerized resources, Thomas Mann (1993) delineates theories of library user behaviour that interact with the services and structures librarians provide. His theories ('Principle of least effort', 'Actual practice' and 'Computer workstation model: part 1 the prospects' and 'Part 2 the qualifications') illustrate the necessity not to think in an either/or way. The 'Principle of least effort' posits that library researchers will 'tend to choose easily available information sources, even when they are objectively of low quality, and, further, will tend to be satisfied with whatever can be found easily in preference to pursuing higher-quality sources whose use would require a greater expenditure of effort' (Mann, 1993). Clifford Stoll (1999) also asserts that laziness and sloppiness are reinforced by digital information sources. Librarians during the CD-ROM period of the mid- to late 1980s certainly saw this behaviour in users and continue to see it now the internet and other digital information sources are widely available. Concerns on teaching the evaluation of information, effective searching using proven search strategies, the rise of plagiarism, using abstracts instead of finding the article in a journal collection, selecting articles based on full-text access rather than intellectual suitability, and the high cost of printing out digital information continue to be common themes in the library and information science literature, electronic mailing list discussions and conference presentations.

A summary of recent arguments from selected electronic mailing lists on the digital versus print debate is provided below (see the Archives of SHARP-L, e-Book-List, JESSE, Libref-L, ARCHIVES@ Listserv.MUOHIO).

Digital: positive points

- Portable: many books can be downloaded at a time
- Efficient: eliminates cost of storage, distribution and transport of physical book
- Waterproof
- Progressive
- Can read in bed without disturbing your partner by turning on a light
- Searchable
- Ergonomically easy to use (no page turning necessary).

Digital: negative points

- Can't curl up in bed with it (take it to the beach, etc.)
- Expensive
- Must teach people how to use
- Unattractive
- Difficult to read
- Hard on the eyes
- Ergonomically hard to use
- Rapidly outdated technology
- Plagiarism easily committed
- Digital Divide – issue of haves and have nots
- Devalues reading culture
- Copyright issues remain unclear

- People print out hard copies anyway
- Preservation problems
- Accuracy – in scanning, items are left out, or decisions are made to exclude special features (e.g. graphics, charts, maps, etc.) found in the print version from the digital version
- Easy to pirate intellectual content.

Print: positive points

- Portable
- Inexpensive
- Attractive, allows for greater aesthetic diversity and creativity
- Can clip a reading light on it and read without disturbing others
- Place can be easily marked/held
- Has worked for over 500 years so why fix it?

Print: negative points

- Old-fashioned
- Yellows, fades, deteriorates
- Rips
- Difficult to search (must have an index to find anything)
- Easy to steal physical item.

As voluminous and passionate as the arguments from electronic mailing list discussions are, none is especially compelling. Interestingly, for both formats the arguments can overlap or contradict, as can be seen above. Experimentation with digital collections has resulted in some anecdotal evidence that supports and refutes the pros and cons for both formats. For example, seven dental schools in the USA required students to purchase a DVD containing the entire four-year dentistry curriculum (Misek 2001; Guernsey, 2000; Alroy, 2000). Dental students at the University at Buffalo, State University of New York, soundly criticized the DVD for its high cost, inefficiency, and support of only the Macintosh platform and not PC (McGarry, 2001). Corporate press releases obviously report rosier experiences with e-books, and the popular press regularly writes of new advances with digital reading matter, such as wine lists for restaurants (Fabricant, 2002) and lighter, handheld devices for travellers (Tedeschi, 2002). Discussions of the traditional book as obsolete (Max, 1994) make for provocative reading, but the greatest likelihood is that reading material will remain in various paper forms (tabloid, glossy magazine, paperback, hardback, oversize, etc.) as well as in electronic forms to be read on handheld devices or computer monitors.

Mann's (1993) analysis of the digital versus print argument sees it as consisting of two propositions:

• Computers in libraries are making vocabulary control and authority work unnecessary.
• Classification schemes for printed books will be rendered obsolete in the digital library.

That is, 'the retrieval capability of the computer is assumed to be so powerful that any records within it – whether catalog surrogates for full texts or the texts themselves – no longer require the standardization, categorization, and integrative relational linkages provided by the work of professional library catalogers' (Mann, 1993). As drawbacks to the overly optimistic and exaggerated claims of digital information, Mann cites the psychological research that confirms difficulty in sorting through too many computer screen display options, preservation problems (machine obsolescence; life span of optical discs, computer tapes or other forms of digital information; authenticity; authority) and costs factors.

Donald Hawkins acknowledges the attractiveness of books and their simplistic engineering which has allowed them to endure. He reminds us that paper's reflective surface properties are conducive to reading in a variety of light sources, whether natural or man made, that books are easy to use, portable and are 'relatively cost-effective'.

But, as a proponent of the e-book, he cites the costliness of producing, storing, shipping and selling traditional books; the difficulty in updating the print resources; and retrieval that is dependent upon human-made indexes (Hawkins, 2000).

Independent publishers question the opportunities for artistic expression in a digital environment that is heavily corporate, profit driven and interested in bestsellers or blockbuster hits. Kathleen Masterson (2000) states that the 'biggest danger to independent literary publishers is that in their anxiety to avoid losing out on the "digital future" they will ape the efforts of the commercial publishers to such an extent that they lose the traits that distinguish them'. Outlets for new talent, experimental forms, controversial or iconoclastic thought may constrict as the environment becomes more difficult and expensive for independent publishers. John Oakes of Four Walls and Eight Windows, an independent publishing company, questions the cost involved in digital publishing and states:

> No publisher should have to pay for the privilege of going electronic. If you can do it for free, fine, but . . . the companies that want to charge you for electronic conversion . . . . Gently, firmly get them out of your life. My proposed titles for the book about this whole fiasco? Here are a few: *Tyranny of the Experts*, *The Emperor Has No Clothes*, or, finally, *More Boys with Toys*. (Oakes, 2000)

The internet and reading

The advent of the world wide web and its pervasiveness in middle-class homes around the world have raised concerns about reading and the traditional book. Readers, though, have taken to the internet and created ways to exchange information about books. Book clubs, online bookstores, library readers' advisory sites, and publishers providing first chapters of books supplement and enhance the print culture – they do not replace it, as was touted by the enthusiasts or warned by the alarmists would happen.

Project Gutenberg celebrated its 30th anniversary in 2001 and its 5000th e-text in April 2002. The availability of public domain literature, free, through a computer, has allowed for access to some of the world's finest classic literature. Yet it has not destroyed the appetite for those same titles in paper form. Although some may download *Pride and prejudice* or *Moby-Dick* and print out these works, paperback copies of these and other classic titles are still widely sold for educational and recreational reading.

To be infatuated with technology and intoxicated with the promises of the digital revolution is to ignore a number of significant issues that are either not being addressed or being addressed ineffectively. These include:

- the digital divide and how it disadvantages large numbers of citizens
- gender bias in the technology
- health and environmental issues
- preservation
- access versus ownership.

The digital divide

Discussion of haves and have nots with regard to technology has appeared in the library and information science literature frequently, and too often at an unsophisticated level of concern. The discussion tends to centre around access to hardware and software, ignoring quality of experiences and equal opportunities. In the print versus digital debate it is frequently stated that the poor and disenfranchised will not have access to the computer technology. But market values do allow for the later penetration of less affluent homes and computer access. For example, in the USA the Census Bureau reports 'a ratio of 9-in-10 school-age children (6-to-17 years old) had access to a computer in 2000, with 4-in-5 using a computer at school and 2-in-3 with one at home . . . [and] 54 million households, or 51 per cent, had one or more computers in the home in August 2000, up from 42 per cent in December 1998' (US Department of Commerce, 2001). The number of homes with computers increased from 24.1% in 1994 to 42% in 1998 (Percent of US households, 2001) But this access to equipment says nothing of educational quality and opportunity, for there is also the social stratification that schools re-create. Studies of educational social stratification suggest that there is a clear difference in the type of teacher and curriculum reserved for the elites. Elites have better-prepared teachers, more resources (print and electronic) and a curriculum which stresses creativity, abstraction, flexibility, independent thought and liberation (Anyon, 1997).

Gender issues

Enthusiasm for the information age and its digital products concerns feminist critics. Suzanne Hildenbrand (1999) presents evidence that historic gender inequities in libraries may intensify with the uncritical embracing of digital products at the expense of traditional resources. Cognizant that questioning the social and political costs of digital information opens one up to dismissal as a technophobe, her research and use of feminist theory nevertheless provide a foundation for understanding job stratification in libraries as digital materials arrive. The network maintenance of e-books and other digital materials has created high salaried positions for usually un-credentialed or less educated males in institutions where women with higher degrees and greater experience are passed over for such positions and/or not rewarded with commensurate salaries. Hildenbrand (1999) and Schiller (1999) critique the information age for its assault on the public sector and government regulations and making a hostile environment for equity issues. Baker (2001), in his chapter 'Thugs and pansies', suggests that machismo, or at least insecurity about the masculinity of male librarians, unduly influences a rush to digital materials, especially at the expense of traditional materials. This association of technology with maleness, and the anxiety surrounding gender roles and sexual preferences, displaces rational reasoning for purchasing certain materials and affects readers and library users by causing resources to be spent in ways that are not always in the users' best interest. Viewing the digital versus print debate through the lens of gender provides a critical analysis of who benefits and who does not, as well as giving a provocative framework for understanding librarianship, publishing and the distribution of information.

Health and environmental concerns

Book readers are stereotyped as wearing corrective lenses, and some such lenses are called 'reading glasses'. Eyesight, print and reading have a long associative history, and the digital environment has only exacerbated the vision and reading issue. Visual ergonomics and vision health management study the proper distance from a computer screen, development of computer glasses and forms of vision therapy needed because of increased exposure to computer use (Anshel, 1994; Thomson, 2002). In the digital versus print argument ergonomics, carpel-tunnel syndrome and vision health are new areas of concern because of the unforeseen overuse of digital reading matter.

The disposal of computers, personal digital assistants and e-books is causing environmental problems (Martinez, 2002). Massachusetts and California have banned the dumping of computer equipment in landfills because of the lead, mercury and other toxic materials used to produce computer technology. The European Union is developing legislation that will make computer producers responsible for taking back their old products, but it is facing opposition from US trade associations such as the American Electronics Association (Silicon Valley Toxics Coalition, 1999). 'Exporting harm: the high-tech trashing of Asia' (Puckett and Smith, 2002) documents the exporting of e-waste and the damage inflicted on poor peoples. Frequently thought of as 'green and clean', digital materials are exposed as toxic, harmful and problematic. The disposal of e-materials also pits developed nations against developing nations. Print disposal does not pose such serious environmental concerns.

Preservation

Digital information, offered as an antidote to the limitations of paper regarding preservation, has its own preservation issues. When purchasing e-journals, digital indexes and e-books the question of ownership arises. Librarians have had to negotiate contract agreements spelling out the rights that the library has to the information, whether it can be archived and whether accessed retrospectively. Machine obsolescence often renders digital information unreadable. This has been recently shown by the obsolescence of the 1986 electronic version of the Domesday Book (McKie and Thorpe, 2002) – the £2.5 million BBC production of a modern-day Domesday Book is inaccessible and unreadable, although the original 1086 record remains readable. Archivists are finding they need new definitions for original document in the digital age. A printout of a digitally stored document does not indicate source or represent changes, and it cannot be authenticated easily.

Access versus ownership

During the 1980s increased spending on technology to provide access to materials saw the argument of access versus ownership arise. Acquisition budgets for purchasing books and serials were raided for purchasing computers, chairs, computer workstations, printers and other hardware to support access to digital collections. Library collections were viewed as becoming less unique and more homogenized as less money was spent on physical collections in order to support the hardware, personnel and subscription prices for virtual libraries. Healy (1995) summarizes the questions of access versus ownership as: 'do librarians rationalize services (e.g. special collections, outreach programs) to ensure that needed material continues to be available on the shelves, or do they sacrifice some of the physical collection and provide access to others' collections via the new technologies?' The definition of a library as a collection of books changes in the digital world because ownership of digital information is a more slippery concept. Intellectual property issues such as copyright piracy, definitions of fair use, authorship, archiving and rights to retrospective collections are all being revised and reinterpreted with controversial results (Electronic Frontier Foundation, 2002).

Conclusion

The dot.com bust and bankruptcies of e-book companies such as netLibrary (Hane, 2002; Rogers and Albanese, 2001; Stutzman, 2001) serve as reminders to be cautious and sceptical when faced with grandiose claims. The demand for books on computer screens has been lukewarm at best – evidence of this may be seen in the fact that the International eBook Award Foundation has discontinued its awards and suspended its activities (Kirkpatrick, 2002). Although the South African Manu Herbstein won the Commonwealth Writers Prize 2002 for his e-book, *Ama, a story of the Atlantic slave trade*, and the American Historical Association offers the Gutenberg e-prize to encourage scholarly writing in the digital environment (Darnton, 2002), and other such prizes exist, the e-book market and acceptance of the medium remain in the infancy stage. Also, netLibrary has faced the reality that individuals do not want their books, and they are concentrating on selling to the library market (Guernsey, 2002). The premature stage of e-books was shown with Stephen King's e-book, *Riding the bullet*. Not only was this book a dismal failure with buyers and readers, but there was the embarrassment of the compatibility issue – Stephen King could not even read his own e-book on his computer (Breitzer, 2000). Computer vendors are attempting to deal with these issues, summed up thus by Calvin Reid (2002): 'As if trying to convince consumers to read books on computers wasn't tough enough, publishers and retailers are also faced with a potentially confusing array of the differing software formats needed to read an e-book in the first place.'

Those who favour digital forms of reading and electronic gadgets should feel safe in the knowledge that new products will be developed for their consumption, but they should also respect the need to maintain traditions and allow for paper formats without dismissing traditionalists as Luddites. Those who favour the traditions must remain vigilant, sift through the hype and apply analysis and criticism to the overstated technology pronouncements. Readers and library users should be able to trust librarians and information professionals to make just decisions, protect cultural heritage, collect and preserve forms of communication and human creativity, and not allow commercial interests to dictate the best ways to perform these tasks.

References

Alroy, D. (2000) Re: Bookbag of the future (March 2), Letter, *New York Times*, (16 March), G10.

Anshel, J. (1994) Vision health management: visual ergonomics in the workplace, *Occupational Hazards*, **56** (4), 55.

Anyon, J. (1997) *Ghetto schooling*, New York: Teachers College Press, Columbia University.

Baker, N. (2001) *Double fold*, New York: Random House.

Benedon, W. (1978) Records Management. In *Encyclopedia of Library and Information Science*, vol. 25, New York: Marcel Dekker, 124–5.

Breitzer, F. (2000) Judging e-books by their covers: format still has hurdles to overcome, *Macworld*, (July), 26.

Commonwealth Book Prize (2002) *Guardian*, (25 April), home pages, 6.

Cox, R. (2000) The great newspaper caper, *First Monday*, **5** (12), available at
www.firstmonday.dk/issues
(accessed 1 May 2002)

Darnton, R. (2002) *What is the Gutenberg-e Program?*, available at
www.theaha/org/prizes/gutenberg/

Eisenstein, E. L. (2002) An unacknowledged revolution revisited, *American Historical Review*, **107** (1), (February), 87–105.

Electronic Frontier Foundation (2002) Unintended consequences:
   three years under the DMCA (Digital Millennium Copyright Act,
   United States), version 1.0 (3 May), available at
   www.eff.org
   (accessed 1 May 2002)
Fabricant, F. (2002) A little help with the wine list: an e-book at the
   table, *New York Times*, (24 April), Sec. F, 2.
Grafton, A. (2002) How revolutionary was the print revolution?,
   American Historical Review, **107** (1), (February), 84–6.
Greco, A. (1997) *The book publishing industry*, Boston: Allyn and
   Bacon.
Guernsey, L. (2000) Bookbag of the future: dental schools stuff 4
   years' worth of manuals and books into 1 DVD, *New York Times*,
   (2 March), Sec. G, 1.
Guernsey, L. (2002) In lean times, e-books find a friend: libraries,
   New York Times, (21 March), Sec. D, 3.
Hane, P. (2002) OCLC completes netLibrary acquisition, raises e-
   book fees, *Information Today*, **19** (3), (March), 17–18.
Hawkins, D. T. (2000) Electronic books: a major publishing revolu-
   tion, Parts 1 and 2, *Online*, **24** (July/August),14–28; **24**
   (September/October), 18–36.
Healy, M. (1995) Information – Access or Ownership?, *ThE-Journal*,
   (November), available at
   http://infotrain.magill.unisa.edu.au/epub/Special_Editions/
   obrit_kents/ articls2.htm
   (accessed 1 May 2002)
Hildenbrand, S. (1999) The information age vs. gender equity,
   Library Journal, (15 April) , 44–7.
Johns, A. (2002) How to acknowledge a revolution: reply, *American
   Historical Review*, **107** (1), 106–28.
Kirkpatrick, D. D. (2002) E-books: an idea whose time hasn't come,
   New York Times, (22 April), Sec. C, 8.
McGarry, B. (2001) Move to electronic texts prompts dental student
   frustration, *Spectrum*, **51** (23), (22 October), 1, 8, available at

http://wings.buffalo.edu/publications/spectrum
(accessed 1 May 2002)

McKie, R. and Thorpe, V. (2002) Digital Domesday Book lasts 15 years, not 1,000, *Observer*, (3 March), 7.

Mann, T. (1993) *Library research models*, New York: Oxford University Press.

Martinez, A. (2002) Old personal computers never die; they just fade into deep storage, *New York Times*, (11 May), Sec. A, 28.

Masterson, K. (2000) *E-books and new digital technology: friend or foe of independent literary publishing?*, available at www.clmp.org/resources/epub.html (accessed 1 May 2002)

Max, D. T. (1994) The end of the book?, *The Atlantic Monthly*, **274** (3), (September), 61–71.

Misek, M. (2001) Universities cut their teeth on DVD, *Emedia*, **14** (3), (March), 32–7.

Oakes, J. (2000) E-books and new digital technology: friend or foe of independent literary publishing?, available at www.clmp.org/resources/epub.html (accessed 1 May 2002)

Percent of U.S. households with a computer, by selected characteristics, 1994, 1998 (2001) *World almanac*, New York: World Almanac Books, 571.

Puckett, J. and Smith, T. (eds) (2002) *Exporting harm: the high-tech trashing of Asia*, San José, California, Basel Action Network and Silicon Valley Toxics Coalition, available at www.svtc.org/cleancc/pubs/technotrash_v3.pdf (accessed 1 May 2002)

Reid, C. (2002) Vendors debate e-book formats, *Publishers Weekly*, **249** (11), (18 March), 31.

Rogers, M. and Albanese, A. (2001) NetLibrary goes on the block, *Library Journal*, (15 November), 16–17.

Schiller, D. (1999) *Digital capitalism*, Cambridge, MA: MIT Press.

Schwartz, L.S. (1996) *Ruined by reading*, Boston: Beacon Press.

Silicon Valley Toxics Coalition (SVTC) (1999) *just say no to e-waste: background document on hazards and waste from computers*, available at www.svtc.org
(accessed 1 May 2002)

Stoll, C. (1999) *High-tech heretic*, New York: Anchor Books.

Stutzman, E. (2001) NetLibrary bankruptcy plan a go, *Camera Business Writer*, (13 October), available at www.thedailycamera.com/business/tech
(accessed 1 May 2002)

Tedeschi, B. (2002) Lighter loads for traveling readers, *New York Times*, (5 May), Sec. 5, 8.

Thomson, I. (2002) Ergonomics, *PC Magazine* (UK), **11**, (March), 80.

Toth, S. A. (1991) *Reading rooms*, New York: Doubleday.

USA. Department of Commerce. Bureau of the Census (2001) Trends: 9-in-10 school age children have computer access; internet use pervasive, Census Bureau report, Press Release, 6 September; Database: A Matter of Fact, entry # MOF91001159.

Part 2

INSTITUTIONAL MODELS AND FINANCE

3

New models of the library in a digital era

Peter Brophy

Introduction

Libraries are continuing to experience a period of rapid change as they reposition themselves in a world where networked services have become the norm. The traditional models on which they have depended, whether these emphasized either the 'warehouse of knowledge' or 'access to information' concept, are becoming less adequate to guide thinking and development in the current era. Coherent models are needed to help libraries find their role and function in a very different environment from that in which they were established.

A particularly striking aspect of the current environment is the number of competitors that libraries now have to face. In the high street bookshops have undergone a renaissance, and book buying is more popular than ever. Publishers are targeting end-users with their products, so that in the academic sector as often as not the user will access a scientific paper directly from the publisher's website. The librarian's expertise in information retrieval is being sidelined by ever-more sophisticated online search tools like Google. Public sector organizations are finding that they no longer need libraries to disseminate their information. All kinds of brokers and information facilitators are finding roles which impinge on those traditionally played by libraries.

In this situation it is inevitable that funders, commentators and other stakeholders are beginning to ask, 'Do we really need traditional libraries?' 'Are they not an expensive irrelevance in this digital era?' 'What do they offer that cannot be achieved more cheaply and more effectively elsewhere?' The 'library' word will no doubt live on, but will the service it represents be recognizable? Sometimes it may seem that the barbarians are already at the gates. Cochrane (1999) gave one vision of this brave new world when he wrote:

> Five years ago, the library at my laboratory used to occupy several large rooms and employ 30 people. It has been replaced by a digital library that is now ten times bigger – and growing fast. This digital library is staffed by only 12 of the original librarians who are now amongst the best html programmers in the company. This digital library has become an essential part of our lives and the work output has gone up tenfold in 10 years.

Is excellence in HTML programming then to become the touchstone of professional excellence? Is this really the future of the libraries that have been developed and cherished over so many years? In order to counter such notions and, more importantly, to take a measure of control of our future, we need well-articulated and robust models of what libraries are and what they can be in the digital age. This chapter is intended to explore some of this territory.

Models of the library: background

Wilson (1999) has suggested that 'a model may be described as a framework for thinking about a problem'. Models are abstractions; they do not attempt to describe particular cases but enable generic features to be identified and analysed. Wilson goes on to remark that more advanced models attempt to describe 'the relationships between theoretical propositions'. In other words, an advanced model helps us to think about the ways in which internal and external interactions occur, or might occur, within the broad environment of the subject under review.

Because the aim of this chapter is to review some of the models of libraries which have been developed, and to begin the process of building on them, it is useful to start by exploring briefly the ways in which thinking about libraries has developed over the years. Any number of starting points could be chosen, but it is perhaps helpful, as a reminder of where we have moved from, to recall the definition of the research library offered by Fremont Rider nearly 60 years ago. It is:

> a vast aggregation of all sorts of book and periodical and manuscript materials, assembled together, not for sustained, or for pleasurable, reading . . . but for 'research', that is, for the purposes of scholarly investigation. Research libraries are, primarily, the stored-up knowledge of the race, warehouses of fact and surmise.
>
> (Rider, 1944)

Or one could, perhaps more profitably, go back even further and recall Ranganathan's Five Laws of Library Science, of which all but the last retain a high degree of relevance (save the somewhat limited emphasis on 'the book'):

- Books are for use.
- Every reader his book.
- Every book its reader.
- Save the time of the reader.
- A library is a growing organism.

These two approaches represent the polarization that for many years characterized debate in the field. Some have seen the core characteristic of the library in the 'holdings' view, with a heavy emphasis on the organization and preservation of humankind's accumulated knowledge. Others have understood the primary role as one of facilitating access to knowledge, and, despite the chronology of the above two quotations, it is this latter view that has become prominent in recent years. Steele (1995) and Sack (1986), among others, have referred to this as a shift towards a Copernican view of the library:

> instead of the traditional 'Ptolemaic' view of the library world with the library at the centre and users at the periphery we now have a 'Copernican' view with the user at the centre and a variety of services and people surrounding and supporting the user.
>
> (Steele, 1995).

Building on this user-centred understanding, some commentators prefer to see the user as representative of the social system of which libraries are a part, and libraries as part of the social 'glue' which holds communities together. Shera, an immensely influential librarian in the USA, wrote:

> A library is . . . a system designed to preserve and facilitate the use of graphic records. It is a social system created to form a link in the communication system that is essential to any society or culture . . . . Its fundamental concern is with the communication of knowledge, ideas, thought.   (Shera, 1980)

More recently, the library as an agent of social change and a bastion against social exclusion has become a common theme. So Linley and Usherwood (1998) wrote:

> Public libraries promote social cohesion and community confidence by fostering connections between groups and communities. Public libraries are seen as community landmarks that reinforce community identity. The library can help individuals, especially older people, overcome the problems of social isolation and loneliness.

The question that arises is how these very different models may be reconciled and given coherence. Libraries can sometimes fall into the trap of being all things to all people. As Kinnell and Sturges (1996) remarked of public libraries:

> At the heart of the issues surrounding public libraries, from their inception up to the present day, has been an imprecise formal definition of their role. Should they be a medium of education and instruction, an information source, a cultural focus for communities, or an addition to people's leisure pursuits through the lending of fiction . . . ? There is so much that [public] libraries do and so much that they could do.

The idea of the hybrid library

The impact of information technology on libraries has been profound. It is, after all, technology for processing and transmitting information, so libraries could hardly stand aside from it. Many have become enthusiastic and expert proponents for the value of IT- or ICT- (information and communications technology) based services. While some have argued for the death of the traditional library, the general consensus is that this is still a long way off – if it ever happens. Zick, for example, has pointed out that software agents (cyber-librarians?) suffer from a number of disadvantages and defects, of which 'brittleness' is a good example.

> Intelligent software programs all face the brittleness problem: how well do they function at a task outside their limited domain? They may excel at chess but flail when queried for obvious information. Librarians are rarely brittle . . . our flexibility from client request to client request is often staggering.   (Zick, 2000)

In the UK the debate in the last few years has tended to centre around the notion of the 'hybrid library'. Rusbridge's is the classic definition:

> The hybrid library was designed to bring a range of technologies from different sources together in the context of a working library, and also to begin to explore integrated systems and services in both the electronic and print environments. The hybrid library should integrate access to all . . . kinds of resources . . . using different technologies from the digital library world, and across different media.
>
> (Rusbridge, 1998)

Brophy and Fisher (1998) pointed out that the basis for this idea lies in the observation that neither purely traditional nor purely digital libraries are able to offer an adequate service, since each suffers from severe disadvantages.

In the traditional model

- each item (book or journal issue, volume when bound) must be used serially (i.e. one user at a time).
- libraries can only stock a restricted range of all the items of potential interest to their users
- publication processes, involving long lead times between author, publisher, printer, distributor and library, mean material is dated even when added to stock
- the cost of stocking little-used items is very high, since staffing and space costs are dominant
- there are high costs associated with handling physical objects (e.g. re-shelving, shelf tidying)
- heavily used items, such as core text books, wear out.

In the electronic model

- the quality of sources is often uncertain or simply unknown
- browsing is difficult at the detailed item level, since computer displays are entirely page-oriented
- the economic model is uncertain, resulting in severe restrictions on

accessing valuable content where suppliers must ensure that copies
cannot 'leak' into general circulation
- there is no consensus on achieving preservation, and no provisions
for legal deposit
- the library is poor at encouraging social interaction, since 'group'
study via technology is as yet artificial, limited and generally unattractive.

They concluded:

> It seems highly unlikely that the undoubted benefits of the electronic
> or digital model will enable it to become dominant in the foreseeable
> future. Not only does the traditional, largely print-based model have
> several advantages, but there is an enormous investment in 'legacy sys-
> tems' – content and infrastructure, including systems for publishing
> content in traditional forms, which has been built up over many years
> and which retains immense value.
>
> (Brophy and Fisher, 1998)

In passing it may be observed that, in common with many debates,
the focus of the hybrid library concept has tended to be found more
in the 'information content' strand of thinking than in concern with
social development. Some, particularly among public librarians, have
countered this by pointing to the library as a non-threatening and
accessible service which offers opportunities to counter social depri-
vation and exclusion. This is an issue considered further below.

The influence of the digital library

Although many different meanings have been applied to the term
'digital library', usually without a great deal of consideration, there
has been a small number of influential studies which have sought to
establish broadly based models suitable for development as the basis
for widespread and achievable library services. Three are described

here, followed by a brief discussion of the significance of portal developments.

Knowledge models

A study by Owen and Wiercx of NBBI (Nederlands Bureau voor Bibliotheek en Informatiewezen), undertaken as a Supporting Study within the European Commission's Telematics for Libraries Programme, developed what were called 'knowledge models for networked library services' (Owen and Wiercx, 1996). The key idea was that 'libraries, as a component of the information chain, act as a link between knowledge sources and users'. The authors suggested that they can therefore best be understood as 'knowledge mediators'. In the context of networked information sources, libraries will no longer be 'restricted to the catalogue' but will make use of a wider range of tools in fulfilling this function. Three fundamental functions of the digital library were defined in this work:

- making available various types of knowledge resources
- providing resource discovery mechanisms which allow users to identify relevant or requested resources and their locations
- providing mechanisms for delivery of specific resources to the user, delivery including both obtaining a resource when it is not already available in the library and passing it on to the user in a suitable way.

Owen and Wiercx developed their model further by examining parallels with traditional library functions (such as user support) and then proposed a series of application models to help libraries incorporate networked resources alongside traditional services. The importance of this model is that it places libraries firmly in the role of intermediaries between the vast and ever-growing wealth of information resource in the world and the individual user seeking a very small, yet to him or her vital, subset of that information. Facilitating access – possibly without even 'owning' any information – is seen as the key.

The MODELS information architecture

The Electronic Libraries Programme (eLib), Moving to Distributed Environments for Library Services or MODELS Project, developed a MODELS Information Architecture (MIA) as a way of describing systems which unify access to service providers such as publishers through an intermediary, while providing flexibility of data presentation to the user – and where the 'user' may in fact be software that processes, analyses and possibly re-uses results in some way on behalf of the human end-user(s). The MIA has been described as both 'a conceptual, heuristic tool for the library community' and 'a tool to assist developers as they think about future systems work' (Dempsey, Russell and Murray, 1999).

In summary the MIA characterizes libraries as brokers which both hide the complexities and differences of underlying resource discovery services and facilitate data flows to enable processes to be automated. The broker is a 'trading place . . . where service requests and service providers come together'. A generalized description of such services includes the provision of:

- user access, including the presentation of an 'information landscape' and support for user profiles
- an applications framework consisting of software and data needed to manage the services, passing data between functions
- distributed service interfaces, which determine and control how requests are presented to underlying services
- access control, including the authentication of users and commercial transactions such as payments (Dempsey, Russell and Murray, 1999).

New or revised underlying services are handled by the applications framework and distributed service interfaces without requiring changes to the user access layer, since the service must operate in an environment of rapidly changing target services. Thus adding a new service should be cost effective (and the library both scalable and sustainable), since it does not require a new user interface to be built.

The applications framework can be defined in terms of four key functions (similar to Owen and Wiercx's analysis described above): resource discovery, location, request and delivery. These require descriptions of the underlying services, including collection descriptions and interface descriptions (i.e. what information is available and the protocols needed to access it), and profiles of users which enable the system to determine access rights, preferences and so on.

Generic intermediaries

Brophy (2000) has published a generic model based on a detailed analysis of library roles and functions across the different sectors. Again, there is stress on the library as an intermediary, but a feature of this model is its emphasis on the need for libraries to maintain sophisticated descriptions both of selected resources and of users. The role is then seen as linking dynamically between the user, based on a rich, personalized profile, and information resources, based on equally sophisticated resource descriptions. Associated with these processes is a variety of support facilities, including those for the encouragement of information skills development and for reference/enquiry services. The key features of the digital library landscape are identified as:

- recognition that in the real world the information of interest to digital library users is to be found in a range of heterogeneous databases and collections, physically distributed but connected by electronic networks and containing objects of many different types
- distributed ownership and rights, including complex intellectual property rights

- a need to provide organization, provenance and authority for items and collections
- a wide range of users with a variety of client systems operating within a broad selection of environments and pursuing many different purposes
- a business need to control access to resources although the models, and especially the economic models, are far from clear
- a role, variously defined, for a broker or other intermediary which connects the users to the resources of interest to them
- a range of standards and protocols for describing resources, for encoding them and for delivering them, including for searching and retrieval. Many existing systems cannot inter-operate effectively, and there is thus a pressing need to agree open standards (Brophy, 2001).

Portals

Considerable interest is being expressed in the development of sophisticated portals, which might be seen as encompassing the intermediary roles of digital libraries. It is important to be clear on what is meant by the term 'portal', since it is often used without great precision. The contrast between an internet 'gateway' and 'portal', using definitions that have been developed in the UK academic sector, may make the significance clear.

A *gateway* provides a list of resources, most frequently using a URL as a link, which have been pre-selected to meet the user's likely requirements. When a user explores the gateway and identifies a resource likely to be of interest, a mouse click or equivalent takes the user out of the gateway and into a service based elsewhere. While the browser 'back' button can return the user to the gateway, the information interaction in essence takes place elsewhere.

A *portal* accepts requests from users and itself interrogates information services it believes may hold appropriate resources. It sends queries to those services and accepts result sets. It then processes those result sets (for example, by removing duplicates) and presents them to the user. In essence the user never leaves the portal.

The commonality between the functions of a library described above (for example in terms of resource discovery, location, request and delivery services) and this definition of a portal will be clear. The question may be whether traditional libraries can evolve portal services fast enough, and with sufficient value-added services, to survive and prosper in the networked information world. The variety of experiments with 'MyLibrary' services suggests that some at least are facing up to the challenge (see, for example, Morgan, 1999).

Multiple models

Although each of the models referred to above is useful in elucidating the core role which libraries can play as information intermediaries, it is fairly clear that they are incomplete. In particular, they have nothing to say about the role of the library as 'social glue' or its proactive function among disadvantaged communities. They also underplay the expertise of trained human beings (the 'brittleness' issue referred to earlier). However this does not, of itself, invalidate them. It may be that what we need is not so much one model – whether it addresses technological or social dimensions – but a series of interlinked approaches, each of which may be emphasized to a greater or lesser extent in real-world services. It is therefore suggested that the following four approaches need to be developed.

Model A: The traditional library

This is the library with which we are familiar, but its place and value need to be re-emphasized. There is a real danger that we could lose the value built up over decades and centuries by generations of librarians and communities through over-enthusiastic and misplaced support for an imagined purely digital future. The traditional library offers:

- a physical building which, through its architecture and interior design, is a statement about the values of the institution or community that owns and sponsors the library service
- bookstock, consisting of carefully selected titles which are organized and presented to meet the likely needs of users
- journals and newspapers, enabling users to keep up to date with their areas of interest through a familiar medium and in warm and reasonably comfortable surroundings
- interlibrary loans, extending the reach of the library beyond its own walls and stock
- reference services, designed to provide answers to the questions users pose
- friendly and knowledgeable staff, who are the approachable face of the library
- a welcoming atmosphere, perhaps including a coffee bar, comfortable seating and bright, modern decor.

It is important that we do not lose sight of the advantages of this model of the library for the future. People will still need physical places, they will almost certainly want to use some kind of individual, physical information objects for some purposes, and they will value the expertise of trained and knowledgeable staff who can help them with their enquiries.

Model B: The memory institution

In Europe there is an increasing emphasis on the library as a 'memory institution', a body – along with museums, archives, galleries, etc. – with a responsibility for maintaining humankind's recorded memory. As Dempsey (1999) has written,

> Archives, libraries and museums are memory institutions . . . . Their collections contain the memory of peoples communities, institutions and individuals, the scientific and cultural heritage, and the products

throughout time of our imagination, craft and learning. They join us to our ancestors and are our legacy to future generations. They are used by the child, the scholar, and the citizen, by the business person, the tourist and the learner. These in turn are creating the heritage of the future. Memory institutions contribute directly and indirectly to prosperity through support for learning, commerce, tourism, and personal fulfilment.

In this description we have another model which has a resonance with established library activity, but expressed in a different way and starting to pose some new challenges. While national, major academic and other libraries have taken on the role of preserving (mainly textual) heritage in the past, commonalities with the broader 'memory institution' arena have not been explicit. Emphases on conservation and preservation, and on bibliographic description, will remain important. To these may be added the museum and archive sectors' expertise in interpretation, as well as their complementary viewpoint on description and preservation.

   The challenge of this model, of course, lies in its extension to the world of digital objects. So much of mankind's digital memory is being lost that a new emphasis on concerted efforts to preserve, conserve, interpret and make available the historical record is vital. This will take individual libraries away from their sole concern with their own holdings into community-wide efforts to find collaborative solutions. As this happens, the role of the library will be redefined.

Model C: The learning centre

Governments throughout the world are giving ever-increasing emphasis to education and lifelong learning, in recognition that a skilled and educated workforce is a prerequisite for economic wellbeing and prosperity. Libraries have long recognized that, since information resources are crucial to education, they have a contribution to make. It is unfortunate, however, that their rhetoric is sometimes divorced from the reality of the role that they can realistically play. In the UK, for example, there has been much talk of the public library as a 'street corner university'. Public libraries 'have a key role to play as "street corner universities"' in promoting education as well as in tackling social exclusion by giving 'information to the "have nots" with access to new technology that they would not otherwise have'(Lister, 1999).

Such sound-bites display a view of the library which is not only unrealistic but potentially damaging to the co-operative relationships which offer a real way forward. Broady-Preston and Cox (2000) found some recognition of the need to see the library as simply one part of a much broader alliance when they reported one of their interviewees as saying:

> We certainly no longer claim to be the principal educator of the com-
> mon man, nowadays we tend to think very heavily in terms of partner-
> ship with adult education, the university of the third age and anyone else
> out there who's got an educational objective . . . . One of our strengths
> is being the people who've got the outlets – we've got more buildings
> than anyone else . . . [but] the concept of the street corner university has
> got to be thought of not so much as a building now, but as consortiums
> of like-minded organisations working in partnership, and the library as
> one of the service points for this street corner university service.

The issue is not only one for the public library sector, however, as may be illustrated by reference to a report prepared for the Formative Evaluation of the Distributed National Electronic Resource (EDNER). EDNER is examining in depth the development of national-level higher education information services in the UK. In this project the Centre for Research in Library and Information Management (CER-LIM) at Manchester Metropolitan University is working with educational researchers in Lancaster University's Centre for Studies in Advanced Learning Technology (CSALT). In an early contribution to the project (Goodyear and Jones, 2001) the CSALT team noted, in examining development projects designed to extend the use of networked information resources in learning and teaching, that little explicit attention appeared to have been paid to underlying pedagogical issues. This is a finding that might perhaps be applied more widely to libraries' efforts to become learning centres. CSALT's model suggests that attention needs to be paid to:

(1) pedagogical philosophy, including

- high level pedagogy – the broad approach being taken, such as a belief in a problem-based learning approach
- pedagogical strategy – the planned actions, which enable tutor and learner to share and develop agreement on what is to be done at a general level.
- pedagogical tactics – the detailed level, such as how learners are to be encouraged to participate, how tutors will deal with problems and so on.

(2) an educational setting in which

- tutors set tasks
- students undertake actions
- the actions are often not the same as the tasks set
- there are outcomes – which are sometimes, but not always, learning outcomes.

All of this happens within an educational context, which includes but is not limited to assessment regimes, student tutoring systems and other support, and so on.

Such a brief summary of the framework needed to be serious players in learning hints at the work libraries need to undertake in developing their thinking and their models of service. Added to these issues is the need to explore how the concept of the library as learning centre relates to current developments in the area of virtual learning environments (VLEs) and managed learning environments (MLEs). Furthermore, at a time when educational providers are all required to undergo frequent and rigorous formal inspections, libraries as learning centres must, if they wish to be taken seriously, accept similar levels of professional, education-led review. Clearly there is a great deal of work to be done in this area.

Model D: The library as community resource

The fourth model focuses on the role of the library within its own community. In part this overlaps with the learning centre model. In part it is well-established, separate territory, since community librarianship has a long track record and has made important contributions. Yet a number of developments suggest that it could become more critical. The role may be more important in a networked world in two ways: partly by offering a human presence when so much else is virtual; partly by helping create community 'glue' in both physical and virtual communities. The UK government, among others, has endorsed this role for public libraries in its report, *Libraries for all*:

> For the foreseeable future there will be a need for community buildings which people are comfortable about entering, a common space where they find an appropriate range of resources and support, and where as a community they can exert genuine influence over what happens.
>
> (Great Britain. Department for Culture, Media and Sport, 1999)

But will libraries fill this role? Black and Muddiman (1997) have their doubts:

> In an age when those concerned with communities and their development preach decentralisation, partnership and empowerment, public librarians seem in danger of reverting to a conservative, insular and defensive stance. Like the public library as an institution, they have become in the main passive and reactive. They respond to constraints and contingencies, but fail to articulate a clear and purposeful narra-

tive of their future role. Visions and Utopias they may have, but in the end, these may not be enough to save the public library from a future on the margins of social and community life.

One of the reasons for the limitations of the library's role in this area in the past has been precisely that 'passivity' has been prized. Librarians cling to their notion of objectivity – Foskett's famous *Creed of a librarian: no politics, no religion, no morals* (Foskett, 1962) – and have enormous reluctance to go beyond 'information provision' to the offer of 'advice'. There have been cogent reasons for this reluctance, including fears over liability if the advice is mistaken, but it leads inevitably to a lack of engagement with real community needs. Citizens, particularly those who are disadvantaged, do not want to be referred to a book or even a web page, but want help in understanding the possible answers to their problems and queries and in making informed choices. At the very least librarians will need to form robust alliances with social workers, counsellors and others to provide the kind of integrated service that is needed.

One possible extension to the model of library as community centre is being explored in a new European Commission-funded project, COINE (Cultural Objects in Networked Environments – see www.coine.org), again led by CERLIM. The key idea is that in the past libraries have occupied a position towards the end of the information supply chain. From the authors, works have gone to publishers, to printers, to distributors and book supply agents, and eventually to library shelves. Finally, the end-user has retrieved the information. But the networked world makes it simple for everyone to become a generator of their own information 'packages' – to become an author. What is more natural than that the library should find a role in enabling ordinary citizens to publish their own stories? What could be more emblematic of the library as a true community resource?

Conclusion

The key concern of this chapter has been to suggest the kinds of models of libraries that need to be developed in the new digital age. To date, the most rigorous work has been concerned with digital, or at least hybrid, libraries. Some of the other models have been the subject of much rhetoric but perhaps less rigorous analysis.

This chapter suggests that the profession needs to engage in more thorough and meticulous analysis of the models that underpin library service development and delivery in the digital age. It is particularly important that we develop integrative models that help us plan a secure future for library services in a world of rapid change, yet continuing need.

References

Black, A. and Muddiman, D. (1997) *Understanding community librarianship: the public library in post-modern Britain*, Aldershot: Avebury Press.

Broady-Preston, J. and Cox, A. (2000) The public library as a street corner university: back to the future?, *New Library World*, **101** (1156), 149–60.

Brophy, P. (2000) Towards a generic model of information and library services in the information age, *The Journal of Documentation*, **56** (2), 161–84.

Brophy, P. (2001) *The library in the twenty-first century*, London: Library Association Publishing.

Brophy, P. and Fisher, S. (1998) The hybrid library, *New Review of Information and Library Research*, **4**, 3–15.

Cochrane, P. (1999) What is the future of man, woman and machine?, *RSA Journal*, **2** (4), 64–9.

Cultural Objects in Networked Environments (COINE) Project www.coine.org (accessed 20 February 2002)

Dempsey, L. (1999) Scientific, industrial and cultural heritage: a shared approach, *New Review of Information and Library Research*, **5**, 3–29.

Dempsey, L., Russell, R. and Murray, R. (1999) Utopian place of criticism? Brokering access to network information, *The Journal of Documentation*, **55** (1), 33–70.

Foskett, D. J. (1962) *The creed of a librarian: no politics, no religion, no morals*, London: The Library Association.

Goodyear, P. and Jones, C. (2001) *Pedagogical frameworks for the DNER (the Distributed National Electronic Resource), Deliverable DC1, EDNER Project*, Lancaster: Lancaster University, Centre for Studies in Advanced Learning Technology. Unpublished.

Great Britain. Department for Culture, Media and Sport (1999) *Libraries for all: social inclusion in public libraries*, available at www.culture.gov.uk/heritage/libl/html (accessed 31 October 2001)

Kinnell, M. and Sturges, P. (1996) Introduction. In Kinnell, M. and Sturges, P. (eds), *Continuity and innovation in the public library: the development of a social institution*, London: Library Association Publishing, xiv–xv.

Linley, R. and Usherwood, B. (1998) *New measures for the new library: a social audit of public libraries*, British Library Research and Innovation Reports 89, Sheffield: University of Sheffield, Department of Information Studies.

Lister, D. (1999) Six councils warned their libraries are sub-standard, *The Independent*, (20 February), 8.

Morgan, E. L. (1999) MyLibrary@NCState: the implementation of a user-centred, customizable interface to a library's collection of information resources, available at http://my.lib.ncsu.edu/about/sigir-99/index.html (accessed 18 May 2001)

Owen, J. S. M. and Wiercx, A. (1996) *Knowledge models for networked library services: final report*, Report PROLIB/KMS 16905,

Luxembourg: Office of Official Publications of the European Communities.

Rider, F. (1944) *The scholar and the future of the research libraries*, New York: Hadham Press.

Rusbridge, C. (1998) Towards the hybrid library, *D-Lib Magazine*, (July–August), available at http://mirrored.ukoln.ac.uk/lis-journals/dlib/dlib/dlib/july98/rusbridge/07rusbridge.html (accessed 20 February 2002)

Sack, J. R. (1986) Open systems for open minds: building the library without walls, *College & Research Libraries*, **47** (6), 538.

Shera, J. H. (1980) Librarianship, philosophy of. In Wedgeworth, R. (ed.), *ALA world encyclopedia of library and information services*, Chicago: American Library Association, 314.

Steele, V. (1995) Producing value: a North American perspective on the future of higher education libraries. In Taylor, S. (ed.), *Building Libraries for the Information Age*, York: University of York, Institute of Advanced Architectural Studies, 77–80.

Wilson, T. D. (1999) Models in information behaviour research, *Journal of Documentation*, **55** (3), 249–70.

Zick, L. (2000) The work of information mediators: a comparison of librarians and intelligent software agents, *First Monday*, available at www.firstmonday.org/issues/issue5_5/zick/ (accessed 20 February 2002)

4

The economic opportunities and costs of developing digital libraries and resources

Simon Tanner

Introduction

This chapter explores the core strategic and economic dilemmas faced by librarians in purchasing or creating digital resources and in developing digital libraries. The opportunity cost of digital libraries is large, and with finite, often reducing, budgets the librarian has never faced so many complex and important economic choices. This chapter describes these choices. It shows how various libraries have reacted and what financial and management strategies are available to cope with the implementation of new technologies and the provision of digital information.

The financial dilemma of becoming more digital

The economic value of libraries in their historical context

Libraries have always had an important economic facet to their existence, development and value to society. The 13th-century historian Ibn Hayyan stated that 'whatever book I want to have I can get on loan from any library, while if I wanted to borrow money to buy these books I should find no-one who would lend it to me' (Lerner, 1998).

When the library stored items of unique value that only a few could afford, then its value was easily apparent and the economies of a central repository of knowledge were obvious. With the democratizing effect of new printing technologies in the Gutenberg era came wider access to cheaper books and thus greater levels of private ownership. While this may have presaged modern competition from the publishing industry with libraries for attention and money, at the time it did nothing but enhance the desire for information and increase the value of libraries. In more recent periods there are many examples of great libraries being created through the largess of benefactors or the tax-paying public which conveyed great social and therefore economic benefits to society without necessarily having much consideration to the initial cost-effectiveness of the development. For example, between 1881 and 1917 Andrew Carnegie contributed US$56 million to build 2509 libraries (www.carnegie.org). Between 2000 and 2002 the Gates Library Foundation is spending US$400 million in funding, software and training to wire every public library in the USA to the internet (Tebbetts, 2000).

New market realities

In the current library environment new market economies mean that, even in those few scenarios of generous funding, every last drop of value must be squeezed from the available resources to maintain that funding now and in the future. Senior library managers are confronting difficult decisions on resource allocation and have to contend with the significant issue of opportunity costs in developing digitally based services.

There are at least three cost aspects to the effective utilization of resources in relation to digital information:

- the immediate start-up costs of either creating or purchasing digital content
- the further implementation costs for establishing a digital library

or even just giving basic access to procured resources
- the costs implicit in managing and maintaining a digital resource in the longer term.

Hand in hand with expenditure on resources is the value and benefit derived from the resource itself, how these are offset against costs and whether it will ever be possible to measure whether 'going digital' can be cost effective. Whether there should be intentions to recover costs in their use or to seek profit in the future is a key strategic question that every library manager will have to address in developing digital information resources or digital libraries.

Digital is becoming ubiquitous

Libraries have experienced a shift in focus over the past decade towards digital formats for information resources. Programmes and improved funding for technology-based projects in libraries, such as seen in the UK's eLib programme, have had a marked effect. In a growing number of libraries there is now an attitude that user demands will be met through digital media and electronic dissemination, as much as through paper-based media. Universities are investing in improved teaching and learning resources via digital media. Other institutions, including the British Library and the US Library of Congress, are developing large technical infrastructures. The implementation of technical infrastructure and the improvement of access to the digital world is seen as hugely important for developing regions such as Africa, which lacks overall investment in social economic growth (Shibanda, 2001), and Latin America, where development has been limited by the relatively lower economic status of the library (Rodríguez, 2001) These shifts at the library or institutional level are a reflection of the macro-economics of the global economy, which contains a much larger informational and digital element than ever before. Cronin points out that America's industrial output weighs about the same as it did 100 years ago, even though real Gross

Domestic Product (GDP) is now 20 times greater, thus 'reflecting the higher knowledge content of contemporary goods and services' (Cronin, 1998). The UN Development Programme describes how this is reflected in other countries: 'In 1999, 52% of Malaysia's exports were high-tech, 44% of Costa Rica's, 28% of Mexico's, 26% of the Philippines'. Hubs in India and elsewhere now use the Internet to provide real-time software support, data processing and customer services for clients all over the world' (UNDP, 2001).

Funding for digital programmes

Examination of some currently funded activity in the digital library arena shows the following levels of funding:

- £50 million for digitization from the UK New Opportunities Fund
- CAN$75 million on content digitization programmes in libraries, museums, and archives via the Canadian federal government's Digital Cultural Content Initiative
- more than £1 million on national digitization projects in Denmark in 2002
- more than £1 million start-up costs at the University of Central England for digital library development
- more than US$12 million expenditure over five years on the Library Digital Initiative at Harvard University.

The challenge facing all the recipients of such funding is to transfer projects into programmes and to develop financial models that will sustain services, resources and development into the period after the initial funding has ceased.

All of this funding, which gives the appearance of large amounts of national and institutional investment, should be set against some fundamental realities of market trends. These show that the number of professional librarians and points of service are reducing and that journal prices, for instance, have increased by 55% in the period 1994 to 1999 (Maynard, 2000). From the librarian's perspective there are fewer professionals, less funding in real terms, a huge acceleration in resource and information availability (at a price), plus a higher demand for digital information from the potential user base.

On the other side of this particular coin is the user perspective. Information users are no longer solely reliant on physical storehouses of information to satisfy their desire for information, nor are they as dependent on the mediation skills of a librarian to answer reference enquiries and search for information, or so they believe. The addition of electronic resources and easy access from home and office may seem, from the user perspective, to degrade the need for library professionals.

The economic elements of a digital library

Digital libraries usually exist within the economic frameworks of libraries or their wider organizations. There are relatively few cases of a digital library being created as a completely autonomous economic unit, even though the needs of managing digital content are so great as to be similar to developing from scratch a whole new library. An example of an autonomous unit is the Cervantes Virtual Library at the University of Alicante in Spain, which has been created as a new, separate library from the already existing University Library (http://cervantesvirtual.com/). There are some important new economic issues raised by the existence of digital resources and the digital libraries to deliver them that illustrate the increasingly complex decisions presented to library managers (Houston and Chen, 2000; Hayes, 1999).

The economic advantages of digital information

There are several economic advantages offered by digital information. These include:

- *the communication effect* – an increased amount of information is exchangeable at a reduced cost
- *access* – the digital format allows multiple accesses to a single copy, thus offering economies of scale in access and delivery; remote access to resources offers users the benefit of not having to travel to the resource
- *flexibility* – the same digital resource can be used for multiple purposes and for many different classes of user without the need to multiply the cost of its creation
- *e-commerce* – digital resources enable a more effective match between buyer and seller
- *production costs* – the creation chain for new publications is generally digital first and then output to analogue format; digital information is 'cheaper to produce, store, distribute and reproduce or copy' (Houston and Chen, 2000)
- *volume enhancement* – generally the greater the volume of digital information the greater the value placed upon the resource or service
- *availability of services* – information accessible via a digital library is usually available at all times, without restrictions related to library opening times or available staffing.

The economic disadvantages of digital information

These advantages are countered by disadvantages that are generally rooted in the contract between producer and user becoming less visible and thus harder to maintain, or in the infrastructure cost:

- *Fair pricing* – it is much harder to place a fair or market price upon a digital resource than the equivalent analogue resource; Houston and Chen (2000) claim that there is still not a 'commonly accepted economic model that can accurately and fairly determine either costs or prices for digital information'.
- *Copyright and intellectual property* – there are complex legal issues

for both the producer and library/distributor of digital information. Because copying something digital is almost without cost and the access to large amounts of that information is much easier, the temptation to make unauthorized use of it is much larger. This remains one of the most contentious and costly issues in digital information and remains unresolved.

- *Access versus ownership* – many of the economic models for digital information offer access at a price, but not physical ownership of a copy. This leads to increased costs to the library in tracking accounts and usage, plus the increased risk in future provision should the provider decide to withdraw the resource from the market if it loses financial viability.
- *Authenticity* – users are generally better able to assess whether a paper resource has validity than they are an electronic one. The role of the librarian is essential is assisting assessment of validity; otherwise demand and pricing may be subject to economic values not derived from the authenticity and stability of the product.
- *Preservation* – digital information is at once more robust and more fragile than paper or film equivalents. While exact copies may be made and widely distributed to ensure security of the data stream, maintaining a basic level of access to a digital resource over time is fraught with problems that must be addressed in a lifecycle approach (Beagrie and Jones, 2001). Costs, therefore, become immediately obvious rather than being deferred into the medium term.

Economic driving forces

By combining the economic benefits and drawbacks of digital information some clear driving forces in its provision become apparent. There has been a rapid inflation in the core library operating costs to account for the explosion in digital resources. This is for two reasons: the cost of acquisition has increased as digital is treated as a value-added purchase, and there is increased capital outlay to develop the technical infrastructure for delivery. While some of these inflationary items are offset by savings in physical space or staffing levels, these are usually outweighed by the need to provide additional computer access points and by staff to provide user support for the new resources. The cost of maintenance is very high; the obsolescence period of equipment is very short, and capital expenditure to maintain access and performance is an additional cost on top of library maintenance.

However, the quality of service may be enhanced through providing much wider resources and because those resources are available any time, any place. Some feel that digital information also can enhance equitable access, and libraries now have much more latitude 'to provide the same set of information resources to all of their patrons, regardless of location' (Majka, 2000), whereas previously simple economics would have precluded additional copies of paper resources being purchased for every location.

Another benefit has been the way that libraries have been able to work together for mutual benefit in terms of resource sharing and consortial pricing, both of these economic strategies being made easier in the digital arena. Libraries are thus in a state of flux, carrying many new types of media and providing access via new technologies to disparate resources – this is the hybrid library concept. Because of these economic driving forces, the value and cost-effectiveness of the library in the digital age has become an issue not just of service quality, but also of survival.

Opportunity and strategic costs of digital libraries

In some cases participation in a digitization or digital library programme may become a distraction from core library goals and thus act as an overall cost to the organization that cannot be offset by the short-term benefit of receiving additional funding. All library managers need to consider the strategic implications of the opportunity cost in pursuing a digital strategy. Such considerations may then mitigate overall risk and lead to hybrid service provision that does not privilege the digital resource to the detriment of the remainder of the library service provision.

The digital opportunity costs are the strategic costs implicit in directing resources to a digital activity, thus reducing the finite available resources that may be expended on some other strategic library activity. There may not be a direct financial outlay corresponding to the opportunity cost, but there are larger strategic costs in potentially misdirecting resources to a weaker digital service strategy, for both the individual and the organization.

In the digital world these opportunity costs are extremely difficult to perceive in advance, but the results of resource misallocation are soon apparent and have long-term consequences. For libraries these may be loss or alienation of their user base, especially by providing collections that are either not suitable or not accessible to their users. The opportunity cost of implementing technology could also be very costly to the organization as a whole, where the library is an integral part of the organization's knowledge chain. For instance the excessive expenditure on knowledge management systems without suitable consideration for non-technical supporting strategies (such as adequate training) has sometimes led to many iterations of the technology being invested in, with diminishing returns.

Planning and financial estimating

Planning and financial estimating are important in assessing potential opportunity costs. Library managers are very good at arguing that they are misssing opportunities to serve users by spending time on detailed planning and financial estimation. However, the opposite is usually true. Snyder and Davenport (1997) suggest that there are three main fallacies to be avoided about planning and financial estimating for technology implementations. These fallacies are:

• System quality is all that matters.

- No projections are better than inaccurate projections.
- Skip the planning when you are in a hurry.

These are usually most loudly heard when new funds become available for which libraries can bid in competition with other libraries. Time is short, so the planning is skipped or reduced to the minimum to enable a bid to be raised. Because big technology development and implementation are complex processes, it seems impossible to make accurate financial projections, so often a guess has as much validity. There is also the tendency to think that, if the system is good enough, if the technology is 'cool enough', then everyone will want to use it. This is clearly a fallacy, as discovered by many dot.com enterprises. Without adequate planning, without financial projections based on evidence and without systems being proven vehicles for delivering what users desire, then no amount of funding will make up for the disruption to the organization, the loss of user engagement, the swing away from core service provision and the opportunities lost.

The growth in competition for resources in the form of funding has meant that meeting obligations on time and to budget is critical, not just for the task in hand but for future funding prospects as well. Recent UK research has shown that in the period 1997–2000 'only 32% of archives, 3% of libraries and 30% of museums had not submitted bids for projects worth over £10,000' (Parker et al., 2001). The competition is real, with current and future success dependent on a good track record of visible success and achievement (Deegan, 2001).

The key to planning and financial estimating in the digital arena is to gather evidence, seek co-operation with experienced partners, test through feasibility studies and measure every aspect (especially people and financial factors) as much and as frequently as practicable. This is a conservative strategy, but actually any other strategy would not be focused on improving services or enhancing user experience – it would be research. This is fine for some libraries and carries the remainder forward, but 99% of libraries are about services that link people with appropriate information and resources. For these the value of the service is not defined by the volume of innovative technology in use but by the number of users served, the amount of information available, the cost–benefit of provision and the satisfaction of users, plus many other non-digital factors.

Therefore, to assess the opportunity cost in considering digital endeavours the library manager will have to weigh up a number of factors; in particular the following tools and strategies are suggested:

- *clarity of service goals and objectives* – without this comparisons of strategies are impossible
- *stakeholder studies* – to define the information needs of key stakeholders in the service and to place some quantifiable measure of value against objectives
- *selection of resources* – by pinpointing the resources under consideration for digital provision then the economic factors can be discovered and evaluated
- *feasibility studies* – to test processes, mechanisms, management frameworks and technical specifications; this is a low-cost sampling route to a closer definition of prices and budgetary needs
- *matrix of requirements* – in any technology implementation the drawing up of a matrix with weighted factors enables effective comparison of disparate potential systems and strategies
- *risk assessment* – without an honest and full assessment of the real risks in the project (such as the likelihood of gaining copyright clearance for free) then the project may run into financial difficulties early in its lifetime and never fully recover; risk assessment allows planning to be put in place to mitigate the worst risks
- *infrastructure survey* – can the installed technical infrastructure cope with the expectations from the proposed system or service; for instance, if video streaming is a core element of service provision, is the network able to support a steady data stream for the number of concurrent users expected?

Digital libraries: are they cost effective?

There is little real economic evidence at present to demonstrate that digital libraries are cost effective in real terms. Kenney and Rieger (2000) state that 'cost-recovery solutions have been advanced, but to date there is little hard evidence that they will succeed'. Libraries are typically subsidized in one way or other, and as such the library ethic will naturally be biased towards providing as many services as possible at zero cost at the point of use. The librarian is more likely to be concerned about the possible inequities of access that the digital format might foster through the varied economic status of users than in trying to squeeze additional money from them. There is some evidence that the costs may in some cases be well balanced by the intangible or indirect benefits, especially in national libraries, where the purpose is not to reap returns on investment in terms of money, but in terms of cultural benefits. This is not a new problem for libraries, but one brought into sharp focus by digital provision.

Parsing the costs and benefits of libraries has always been an abstruse exercise. Library costs are more easily determined than real but often unquantifiable benefits. Interesting and useful insights into cost trade-offs can accompany the traditional analysis of costs versus benefits for digital library products. These trade-offs should give library directors and financial managers grounds for reflection as they contemplate their goals and budgets (Majka, 2000).

The route to visible cost-effectiveness, and thus sustainability, is not going to be down a single path, nor will it be a simple process. The hybrid routes both for the future of the library service and for the models of sustainable development seem the most viable. Beagrie (2000) argues that 'digitized cultural information, no matter how worthy, is not in and of itself commercially viable and that the investment in its development, maintenance and delivery can only be achieved through initial non-commercial development funding or sustained by supply of value-added services. Librarians need a means to demonstrate they are delivering benefits, however intangible, to justify the costs being expended in digital libraries.

A heuristic solution for cost-effectiveness

The measurement of benefits versus costs will remain a difficult issue. As digital services are still developing and with benefits often not yet fully achieved, the point of breaking even, let alone showing profit, seems elusive. For managers looking to inform their planning and investment decisions, there are obvious advantages in being able to predict when or whether a digitization or digital library project will make the transition from loss into profit. The terms 'loss' and 'profit' here are not only equated in straightforward cash terms but will also include other, more elusive, factors such as actual service enhancement and improved user access and organizational prestige.

It is important to recognize that the costs of starting digital library projects have many similarities with the costs for developing the traditional library. The costs for digital content are front-loaded with a large, usually irreducible, cost for initial procurement and management, which is then offset against the marginal cost of delivery. Similarly, the cost of storing the digital object is low, but the initial cost of creating the environment in which it can be stored and retrieved is large. There are obvious parallels with traditional libraries: 'once the huge fixed investments [in library buildings] are made, the very small marginal cost of adding a volume to the physical collection supports the continuation of this pattern – until, of course, a library runs out of space, at which time the effort to raise the capital funds begins again' (Guthrie, 2001).

A model for the break-even point in digital library development is illustrated in Figure 4.1 (Deegan and Tanner, 2002). Deegan and Tanner state that, although this is founded on real project information, it remains an hypothesis, as actual development is not yet mature enough to provide firm numeric evidence for the whole model. The model shows that the initial investment in infrastructure is very high

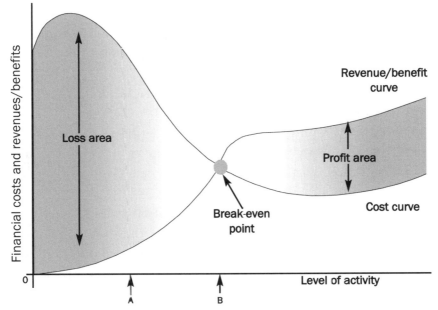

Fig. 4.1 *The break-even point for digital library development*

and that benefit initially is small, but grows quite steeply once initial implementation is completed and services are online. Looking to the future, it seems clear that benefits will eventually plateau and with prudent management grow steadily into the future – there are very few scenarios of constant exponential growth in any sector of activity. Costs and investment to achieve goals will be reduced over time after the initial hump of large-scale, start-up investment in infrastructure, staffing, training and other implementation costs. Deegan and Tanner show the point these two curves cross (marked as B on the level of activity axis) as the break-even point where costs and revenue/benefit are perfectly balanced for a short time. The costs of running the service will not plateau at a relatively low maintenance level, but will eventually start to grow again as technology, equipment and infrastructure need to be renewed and the costs of digital data preservation become evident (Beagrie and Jones , 2001).

The differentials between the cost and revenue/benefit curves are small, but they are not limited; they extend into the future and thus have the potential to repay investment in the medium term. Deegan and Tanner's illustration assumes a zero point starting position, which is unlikely in technically developed organizations. As more technical infrastructure, staff and experience become available, then the starting position will slowly move to the right along the level of activity axis until it reaches the point marked A. This means that 'the initial cost of developing a digital library or digital resource is high at present, but as activity and available experience increase the initial costs will be lower and the benefits will be achieved more quickly' (Deegan and Tanner, 2002).

Each library will have its own measure of costs and benefits and should attempt wherever possible to transfer these into measurable terms that become comparable, whether through weighted ratings or direct cost comparisons. Without these sorts of quantifiable measures, managers leave themselves open to the criticism of not achieving profit or institutional goals, without having satisfactory proof to the contrary. As Guthrie (2001) suggests, 'if it is not possible to monetize these benefits in a situation where the costs and benefits can be estimated and projected with confidence, how are we going to address the huge electronic archiving problem?'

Addressing information inequities via digital resources

As important as it is to 'monetize' the costs and benefits in digital libraries, there are wider international issues that digital resources and libraries can address in terms of information inequalities because of the new pricing structures and opportunities possible in the digital arena. The traditional publishing and distribution mechanisms do not deliver information products in an equitable way, because it is just not practicable for a library in a developing country to afford the same level of information provision as a rich country. For example, whereas a US medical library might subscribe to about 5000 journals, the Nairobi University Medical School Library, long regarded as a flagship centre in East Africa, receives just 20 journals. In Brazzaville,

Congo, the university has only 40 medical books and a dozen journals, all published before 1993 (Witten et al., 2001). Digital resources, by breaking the relationship between the cost of production and distribution from that of intellectual property charges, offer an opportunity to deliver more equitable information services.

Because of the new economic possibilities offered by digital resources, the World Health Organization has been able to negotiate a deal with publishers of approximately 1000 leading medical and scientific journals. This deal offers electronic subscriptions at 'free or deeply-reduced rates to institutions in the developing world' for an initial three-year period (www.who.int/inf-pr-2001/en/pr2001-32.html). It is clear that this is an enormous information benefit to the developing world. It also offers economic benefits to all parties that are possible only in the digital arena. The digital distribution cost to the publishers is virtually nothing, but they gain significant market knowledge and customer loyalty in a new and developing world market. The opportunity for co-operation and central or consortial negotiation enables bodies such as the World Health Organization to predict costs against significant benefits and to manage these in a predictable way with easily apparent value. There is a possible downside to differential or consortia pricing for networked products, particularly if the product or service started to drift back into the home market at reduced rates, if the intellectual property contained in the products were abused or if the home market objected to the lower price being offered to other countries. Educating the market on the producer and consumer side seems the best route to wider understanding and appreciation of the rights and responsibilities inherent in differential or consortia pricing.

The most serious obstacle to these sorts of initiative is the low rate of take-up by the local populace. While the rapid obsolescence of computers in technically developed countries offers a filter-down availability for lower specification machines to developing countries at almost zero cost, the infrastructure to support network access is still very expensive to install and use. There is a lag between availability and access that will be bridged in the coming years by better infrastructure, including global satellite communication networks, but this will take time and money. It is clear that, whatever the barriers now in place, these will be removed over time, and technology will convey global information benefits, as described by the UN Development Programme:

> Information and communications technology can provide rapid, low-cost access to information about almost all areas of human activity. From distance learning in Turkey to long-distance medical diagnosis in the Gambia, to information on market prices of grain in India, the Internet is breaking barriers of geography, making markets more efficient, creating opportunities for income generation and enabling increased local participation.
>
> (UNDP, 2001)

Conclusion

The route to economic sustainability of digital libraries is not going to be down a single path, nor will it be a simple process. The hybrid routes both for the future of library service and for the models of sustainable development seem the most viable. Libraries need to find the balance between their analogue and digital collections and services. Just because something is in digital format, or the service is delivered digitally, does not make it inherently more valuable or indeed more commercially viable. The pressures from publishers and dot.com e-libraries such as Questia (www.questia.com/) stretch the library manager's justifications for digital library development in a particular institution and the need to secure sustainable funding for all information services, whether e-based or traditional. The key remains the unique value of the content that could be presented and preserved and the unique modes of access that may be designed for the future. The justifications for investment are not going to be only financial, and strategists need to seek other tangible advantages for their organization to help justify the initial costs of development.

The future is ripe with possibilities for new fund-raising and charging models, for both publishers and intermediaries, with a constantly expanding potential audience. The rise of micro-payment, especially in mobile telecommunications technologies enabling internet access such as i-Mode and WAP, has been meteoric. Under this scheme the user pays a tiny amount per transaction direct to the phone connection supplier, which appears in the overall connection bill rather than constantly referencing a credit card number or other comparatively cumbersome payment method. With the further rise of WebTV the sheer penetration into the global population of access, at potentially low cost, to digital information resources and services will lead to greater competition for revenue and for an audience in a truly global economy.

The sustainable development of digital libraries requires support at the highest levels of each organization and should be based on long-term goals and returns rather than short-term gains. It is imperative that short-term technical solutions for short-term gains are avoided and that an internet-based system does not risk the underlying valued digital assets by focusing on the immediate problems of funding or technical infrastructure. What appears impossible today might be tractable in the near future, so it may be strategically sensible to wait until the task can be done correctly rather than to jump in too early, wasting resources and potentially having to rework the project because it is in advance of current possibilities. Library managers need to bear in mind the opportunity costs inherent in chasing new technology and ensure they focus on core information goals.

To invest in technology over people, services or other resources is extremely tempting to strategists, governments, funding agencies and senior managers who are driven by some fundamental prerequisites as defined in the 'law of disruption' by Downes and Mui. This states that 'social, political and economic systems change incrementally, but technology changes exponentially' (Downes and Mui, 1998). In other words we will spend more on technology in the hope that society and institutions will eventually catch up with the changes brought about through technology, rather than allow social and economic need to drive technological development. The challenge for librarians is to not expend as much effort in 'catching up' with technology and spend more time in adjusting and taming it for society's information needs.

References

Beagrie, N. (2000) Economy: some models for sustaining innovative content-based service. In *Digitising journals: conference on future strategies for European libraries, Copenhagen, Denmark.*

Beagrie, N. and Jones, M. (2001) *Preservation management of digital materials: a handbook*, London: The British Library.

Biblioteca Virtual Miguel D. Cervantes
　http://cervantesvirtual.com
　(accessed 28 April 2002)

Carnegie Corporation of New York
　www.carnegie.org
　(accessed 28 April 2002)

Cronin B. (1998) Social dimensions of the digital revolution, *Journal of Information, Communication and Library Science*, **4** (4), 3–9.

Deegan, M. (2001) Money makes the world go around: finding further sources of funding. In *Digitisation solutions, HEDS conference*, available at
　http://heds.herts.ac.uk/conf2001/heds2001_deegan.pdf
　(accessed 10 March 2002)

Deegan, M. and Tanner, S. (2002) *Digital futures: strategies for the information age*, London: Library Association Publishing.

Downes, L. and Mui, C. (1998) *Unleashing the killer app: digital strategies for market dominance*, Cambridge, MA: Harvard Business School Press.

Guthrie, K. M. (2001) Archiving in the digital age: there's a will, but is there a way?, *Educause Review*, (November/December), 56–65.

Hayes, R. M. (1999) The economics of digital libraries. In *Simposio Internacional: Impactos das Novas Technologias de Informacao, 23–24 Sept. 1999, University do Sao Paulo, Brasil*, available at www.ime.usp.br/~cesar/simposio99/hayes.htm (accessed 10 March 2002)

Houston, A. L. and Chen, H. (2000) Electronic commerce and digital libraries. In Shaw, M. et al. (eds), *Handbook on electronic commerce*, Berlin: Springer Verlag, 339–63.

Kenney, A. R., and Rieger, O. Y. (eds), (2000) *Moving theory into practice: digital imaging for libraries and archives*, Washington, DC: Research Libraries Group.

Lerner, F. (1998) *The story of libraries from the invention of writing to the computer age*, London: Continuum Publishing.

Majka, D. R. (2000) The 'great exchange': the economic promise and peril of the digital library, *The Bottom Line: Managing Library Finances*, **13** (2), 68–73.

Maynard, S. (2000) *Library and information statistics tables for the UK 2000*, Loughborough: Loughborough University, Library and Information Statistics Unit, available at www.lboro.ac.uk/departments/dils/lisu/list00/list00.html (accessed 10 March 2002)

Parker, S. et al. (2001) *The bidding culture and local government: effects on the development of public libraries, archives and museums*, Newcastle-upon-Tyne: University of Northumbria, School of Information Studies, available at http://is.unn.ac.uk/imri (accessed 10 March 2002)

Questia – The online library of books and journals www.questia.com (accessed 28 April 2002)

Rodríguez, A. (2001) The digital divide: the view from Latin America and the Caribbean. In *67th IFLA Council and General Conference*, available at www.ifla.org/IV/ifla67/papers/111-114e.pdf

(accessed 10 March 2002)

Shibanda, G. G. (2001) Skills and competencies for digital information management in Africa. In *67th IFLA Council and General
Conference*, available at
www.ifla.org/IV/ifla67/papers/009-143e.pdf
(accessed 10 March 2002)

Snyder, H. and Davenport, E. (1997) *Costing and pricing in the digital
age*, London: Library Association Publishing.

Tebbetts, D. R. (2000) The costs of information technology and the
electronic library, *The Electronic Library*, **18** (2), 127–36.

United Nations Development Programme (2001) *Human development
report 2001: making new technologies work for human development*,
Oxford: Oxford University Press.

Witten, I. H. et al. (2001) The promise of digital libraries in developing countries, *Communications of the ACM*, **44** (5), 82–5.

Part 3

BOOKS AND 'READERS'

5

Reading in a digital age

Catherine Sheldrick Ross

Introduction: The death of reading/the physical book/libraries?

At a time when digital forms of information and entertainment are gaining in importance, there is considerable worried talk about the death of reading and the obsolescence of the physical book. In a frequently recalled scene from Victor Hugo's *The hunchback of Notre Dame,* a scholar holds up an early printed book, looks towards Notre Dame Cathedral, and says, 'Ceci tuera cela' – 'This will kill that' – which is to say, the new technology of the book will kill the institution of the church. As linguist and researcher at Xerox Palo Alto Research Center Geoffrey Nunberg (1993) points out, this is an old anxiety that has been voiced repeatedly whenever a new technology has appeared on the scene. Once the new artefact has a toehold, so the technological determinist argument goes, the new will take over completely and drive out the older artefact as well as the institutions and practices that have shaped its use. Now it is the computer and the e-book that are said to be killing libraries, the physical book and reading itself, together with all the values, practices and social relations associated with print culture.

Sounding the alarm, books like Jane Healy's *Endangered minds* (1990) and Sven Birkerts' *Gutenberg elegies* (1994) argue that the capacity for reading sustained text is being threatened by competition from visual media such as television and video and/or by the fragmentary nature of the discontinuous hypertext read on screens. Healy argues that what we now call thinking itself – the ability to pursue the development of an idea, step by step in a logical chain of reasoning, through sentences and paragraphs – is an outgrowth of the linearity of print. In *The Gutenberg elegies: the fate of reading in an electronic age*, Birkerts has made himself an advocate, *par excellence*, for print culture, literary texts and the act of deeply engaged reading. He is at his best when he describes the pleasures of reading codex books: the experience of curling up with a work of literary fiction; the sensory engagement with the physical book, touching its binding, turning its pages, inhaling its smell; the way fiction can draw you into a world so that you take up residence in it and it stays with you long after you close the physical pages of the book. 'What', he asks, 'is the place of reading, and of the reading sensibility, in our culture as it has become?' His answer is that the place is shrinking and the sensibility withering, under attack from electronic media in its myriad forms. For him 'a book is solitude, privacy; it is a way of holding the self apart from the crush of the outer world'.

Birkerts' argument is that people – that is, younger people who have grown up with television and computers – are no longer acquiring the ability to read deeply and enter these private, imagined worlds. As evidence, he points to undergraduates in his course on the American short story who have an attention span so shortened and a sense of time so fragmented that they are unable, or unwilling, to sustain a prolonged engagement with Henry James's 'Brooksmith'. In its contrast of 'deep reading' of real books with the superficial skimming of hypertexts, Birkerts' analysis recalls a similar distinction made by the German historian Rolf Engelsing and frequently cited (e.g. Darnton, 1982). From the Middle Ages until some time towards the end of the 18th century, Engelsing noted, people had just a few books such as the Bible and read them 'intensively', over and over again, often aloud and in groups. Then there was a shift to 'extensive' reading as the presses produced an enormously expanded range and number of publications, especially periodicals and newspapers, but

also novels. After this '*Leserevolution*', people in their pursuit of commodified amusement tended to read a text only once before racing
on to the latest work.

In Engelsing's anxiety about the expansion of extensive reading
and the desacralization of the printed word, we can see some familiar oppositional pairs: deep reading versus superficial reading; active
engagement with a central canonical text versus passive consumption
of a stream of ephemeral materials whose apparent novelty conceals
the fact that they are essentially the same commodified and repeated
product (e.g. newspapers, magazines, dime novels, series books,
genre books such as romances or detective fiction, bestsellers). There
are, it seems, many accounts of the decline of reading from some
golden age of the past when reading was deep, intensive, whole and
life affirming. The exact point when the fall is alleged to have happened, however, differs in these various accounts – the late 18th century with the expansion of printing; the 19th century when the
invention of the steam-powered press made cheap fiction, magazines
and newspapers available to whole classes of readers who had never
before read much of anything; the late 20th century with the global
reach of the internet.

Telling a completely different story about reading and literacy,
cyber-theorists such as George Landow (1992), Richard Lanham
(1993), James O'Donnell (1998) and Janet Murray (1997) agree that
the digital media have introduced a transforming shift in reading and
in the ways in which cultural products are produced, disseminated
and received. However they see exciting new potentials for cultural
expression and for education, as the electronic environment opens
up new spaces for reading and for writing. Murray's book, *Hamlet on
the holodeck*, sketches out the aesthetics of a new type of narrative that
has not yet been invented but can be seen in embryo in the work of
video game designers, computer programmers, and web page designers. She looks forward to cyber storytellers discovering new modes of
representation by taking full advantage of the technology's strengths:
its interactivity; non-linearity; ability to create immersive three-

dimensional landscapes and machine-based fictional characters such as 'chatterbots'; and its convergence of text and images, audio and video. As her title suggests, Murray sees in the immersive electronic environment a powerful technology of sensory illusion that is 'continuous with the larger human traditions of storytelling, stretching from the heroic bards through the nineteenth-century novelists' (Murray, 1997).

It may be helpful to place these anxieties about the death of the printed codex book and of deep reading within the context of other gloomy prognostications – the alphabet and inscription will kill human memory; the book will kill the cathedral; the photograph will kill painting; the telephone will kill the art of letter writing; film will kill live theatre; television will kill film, radio and/or books; interactive computer-based environments such as the video game and the internet will kill broadcast media such as television. These predictions were only partially right. The threatened artefact or technology in most cases has turned out to have its uses and has prevailed in a niche where its utility or pleasure could not be matched by the newer, competing technology. Admittedly there are some e-book enthusiasts who predict a new age of reading right around the corner when printed books will be displaced almost entirely by e-books – just as soon as some little problems are ironed out and a few technological improvements achieved such as screen displays of 300 dpi, portable and inexpensive e-book reading devices that can match such valued features of printed books as allowing for writing marginal notes, and the universal acceptance of a common e-book standard. It is more likely that digital books will take their place alongside printed codex books, and that readers will read both, at any given time choosing the format that suits their purposes, just as people now get news from various media such as radio, television, newspapers and news magazines. (For helpful discussions of this question, see Epstein (2001), Jenkins (1998) and Pang (1998).) An area of research that needs more attention is the study of how readers actually engage with these different formats of digital text vs printed text, their reasons for choosing one format over another, and the different values and satisfactions they assign to reading each.

Fostering reading

Since the ability to read extended complex texts, including texts on screens, is a requirement for full participation as a citizen and worker in a modern economy, understandably there is considerable concern about literacy and the conditions that foster it. Research on reading and literacy is converging on the conclusion that the 'literacy crisis' is not that people are unable to read – in Westernized societies, a higher percentage of people *can* read than ever before. The problem is that too small a proportion of readers read well enough to cope with the complex literacy demands of modern society. Many children, especially boys, choose not to read. Of the poor Grade 4 readers in a Texas school, 40% claimed they would rather clean their rooms than read. One child was even more emphatic: 'I'd rather clean the mold around the bathtub than read' (Adams, 1990).

In *The power of reading* Stephen Krashen (1993) draws together the results of a large number of American research studies on the question of what factors bring about success in reading and writing, and he suggests a remedy. He recommends fostering a certain kind of reading that he calls 'free voluntary reading'. Free voluntary reading 'means putting down a book you don't like and choosing another one instead. It is the kind of reading that highly literate people do obsessively all the time' (Krashen, 1993).

The connection between voluntary reading and powerful literacy is that people learn to read by reading. What keeps children reading for the countless thousands of hours necessary to produce the bulk of reading practice that creates confident readers? The answer seems to be the pleasure in the reading experience itself. It appears that a reader learns to read not by drills and exercises but by reading a lot of text that is meaningful and personally rewarding. Avid readers often describe themselves as coming from households where they were surrounded by books. They claim that as children they had been voracious readers, that they 'read everything' and 'read indiscriminately'. Unlike non-readers, who say they lack the time to read and find reading hard, avid readers say that they can read 'any time' and that they build opportunities for reading into their daily routines. Reading is interwoven into the texture of their lives, not separate from it.

In a study that I conducted based on more than 200 open-ended interviews with pleasure-readers, interviewees were chosen because they said that reading formed a very important part of their lives. These readers turned out to be adept at choosing books that suited their needs and purposes at any given time, depending on the satisfaction that they were looking for (Ross, 2001). They talked about the way that books – often fiction books, but sometimes non-fiction such as biography or travel – gave them a pleasure that they could not find in any other way. And beyond providing pleasure the books helped them in important and diverse ways, but not in a way that an outsider could have predicted from the titles of the books themselves.

Avid readers talked about books that 'opened my eyes' to a new perspective or 'opened a door' on a new reality. They said that a particularly significant book was a model for living – that the narrative representation of human experience within a novel offered examples to follow, rules to live by and sometimes inspiration. In some cases reading changed the readers' beliefs, attitudes or pictures of the world, which change in turn altered the way readers chose to live their lives after the book was closed. Other books reinforced the familiar or confirmed what was already believed.

Often the reader talked about the way an experience in a book seemed resonant with their own experience, claiming that the book 'sounded a chord' or 'struck a key'. Another large group of readers said that the significant book provided reassurance, confirmation of the self or inner strength. For books that offered comfort, especially childhood favourites that were constantly reread, readers used the metaphor of the book as a 'friend' or as 'comfort food'. One reader said, 'Books have different values depending on the stage of your life you're at when you read them. Sometimes your life intersects with a book and you can really benefit from it. . . . . I think that when I approach books I look for how they address my life' (Ross, 1999). It was evident from their description of how books helped them that readers themselves play a crucial role in enlarging the meaning of the text by reading it within the context of their own lives. Through their act of making sense of texts and applying them to their lives, readers creatively rewrite texts.

From their experiences with many kinds of reading materials – from reference works meant to be mined for a particular piece of information to fictions that invite immersion in a world – practised readers discover that there are many kinds of reading and many literacies. They learn that there are many different kinds of texts to be read for different purposes. They know that they need to adapt their reading strategy to the text and the task at hand, sometimes skimming quickly over long expanses of text, using chapter headings, introductions, summaries and captions to help them get the gist and at other times reading slowly, intensely and closely, and perhaps rereading.

In the electronic world of hypertext and web pages, readers are developing new strategies of reading, and researchers are only at the very beginning stages of discovering what those strategies are. In *Cyberliteracy* Laura Gurak (2001) argues that what is needed is a 'critical technological literacy, one that includes performance but also relies heavily on people's ability to understand, criticize, and make judgments about a technology's interactions with, and effects on, culture'.

Reading as a social act

It may seem counterintuitive to claim that reading is a profoundly social activity. Once the technologies of the electric light and cheap books made it possible for ordinary people to have access to private space and plentiful reading material, the predominant image of the reader has been of the solitary individual engaged in an essentially private and unshared activity, somehow removed from the concerns of the public sphere. As Elizabeth Long (1992) points out in a paper in Jonathan Boyarin's edited collection, *The ethnography of reading*, the solitary female reader is a familiar iconographic image.

There is a tension in our view of reading, because time spent reading is time taken away from socializing. Readers like to retreat to what Virginia Woolf has called a 'room of one's own' to have a quiet space for reading (and writing). Unlike reluctant readers who find reading hard work, avid readers who read for pleasure can read anywhere. They carry books with them so that they can read for five or ten minutes while waiting in queues or at the doctor's surgery. But reading in snatches is second best, in comparison with uninterrupted reading. Lynne Schwartz (1996) in *Ruined by reading* mentions the voluptuousness of reading late at night, when reading can be prolonged and uninterrupted. 'Sometimes at the peak of intoxicating pleasures, I am visited by a panic: the phone or doorbell will ring, someone will need me or demand that I do something.' Italo Calvino (1981) begins his splendid novel about reading by advising the reader, 'You are about to begin reading Italo Calvino's new novel, *If on a winter's night a traveller*. Relax. Concentrate. Dispel every other thought. Let the world around you fade. Best to close the door; the TV is always on in the next room. Tell the others right away . . . "I'm reading! I don't want to be disturbed".'

On the other hand, reading is nourished and sustained in community with others. Reading is learned in the social context of family and school; supported by book exchanges and reading materials given as gifts at ritual times such as birthdays; valued in particular ways by cultural authorities, including book reviewers and curriculum committees that draw up university reading lists; and directed toward particular books and authors by practices that determine what reading materials are available to read, including the activities of publishers, booksellers and collection management librarians. Long (1992) claims that 'the ideology of the solitary reader . . . suppresses recognition of the infrastructure of literacy and the social or institutional determinants of what's available to read, what is "worth reading", and how to read it'. Boyarin (1992), reflecting on the papers in his collection, including Long's, says that most of the essays 'share the task of dissolving the stereotype of the isolated individual reader, showing that not only is all reading socially embedded, but indeed a great deal of reading is done in social groups'.

Interested in making visible the reading practices of groups of readers, Long conducted an ethnographic study of reading groups in the 1980s in Houston, Texas. She found that readers, in the very act of joining one particular reading group rather than another, were defining themselves with respect to social and political values. Women who joined reading groups at a time in their lives when they were isolated in the suburbs with young children talked of their reading group experience as providing 'a "lifeline" out of their housebound existence into a world of adult sociability and intellectual conversation' (Long, 1992). Big box bookstores are taking advantage of the social nature of reading by creating spaces for people to congregate, drink café latte, read, attend author readings – and buy books. More recently the internet has become a space that connects readers through virtual book discussion groups, listservs, websites and real-time chat. Perhaps the most celebrated example of the power of electronic connectivity to amplify book group activity has been the Oprah Book Club. Mary Chelton's article (2001), 'When Oprah meets e-mail: virtual book clubs', provides an excellent overview of websites for online book clubs as well as the numerous resources on book discussion groups, how to start and maintain them and suggestions for books to read.

Reading (electronic) text

On 14 March 2000 Stephen King published a novella called *Riding the bullet*, making it available in electronic format only. Readers had to read it using their computers, handheld devices, or dedicated electronic book readers. Over 500,000 people tried to download it within the first week, paying US$2.00 for the privilege. The Barnes and Noble website described the event as 'one of the biggest events in eBook history' and reported that King's publisher, Simon & Schuster, 'is also excited about this new revolution in publishing': 'What's exciting is that we are able to go from Stephen King's computer to the reader in a fraction of the print-publishing arc.' We do not know much yet about the readers' experience with reading e-fiction, although there is accumulating anecdotal evidence that people still tend to print out units of text that are any longer than a few screens (Lynch 2001); that some e-book readers have been frustrated at the way that a 230-page book has to be read as 2300 separate screens on a Palm Reader; but that others find it handy to have a novel on their Personal Digital Assistant (PDA), just in case they are stuck in a queue and need something to read.

Electronic text read from a screen will prevail in cases where the reader needs ready and quick access to relatively short pieces of text, especially where currency is an issue. Not surprisingly, the migration to the electronic world has worked best with textual forms that are not intended to be read in a linear way but instead are consulted, dipped into and read in small chunks – e.g. reference works of all kinds, operating manuals, newspaper articles, websites, and the like. In successful translations to digital formats, such as has happened with the electronic encyclopedia, the genre has been redesigned to exploit the strengths of the digital environment, becoming more like a database that can be searched to deliver up materials relevant to a reader's interest. And if the reference database is the model *par excellence* for the new digital text, it appears that in the electronic environment whole books are being absorbed as elements in a huge database that is to be consulted but not read. Reviewing recent enterprises such as netLibrary (OCLC), Questia Media, and Ebrary.com that are racing to put full texts of hundreds of thousands of copyrighted books, old and new, on the web, Lisa Guernsey claims that 'the new effort to build an electronic library is not about reading at all. It is about the power of electronic searching. With digital scanning, texts of works that may be decades old can be mined for those few morsels of insights that may enhance a research paper or help prove an argument' (Guernsey, 2000). Some academic libraries, initially concerned that a 24-hour borrowing time limit for e-books might be too short, have discovered that the average time that users are spending with a netLibrary e-book is about ten minutes. However, to make sense out of this number we would need to know the average time spent by academic library users with printed books, a statistic that is not available.

Clifford Lynch singles out the scholarly website as an example of a new genre designed to engage readers in new ways. He points out that it

> links and organizes many small chunks of text with mul-
> timedia content and provides the ability to search and
> navigate among them. It may also include interactive
> software components such as simulations, and use the
> communications capabilities of the internet to build an
> interactive community around the work and its subject
> matter.   (Lynch, 2001)

Theorists such as George Landow (1992) and Ilana Snyder (n.d.) see the new genres of hypertext, hypermedia and hyperfiction as providing a concrete illustration of concepts of reading that were first explored by post-modern critical theory. According to hypertext theorists, hypertext challenges our print-based notions of authors, readers and the integrity of the text in the following ways:

* With printed books the author determines a fixed order in which ideas are presented and read, whereas with hypertext the author presents options, but the reader chooses which links to pursue. The reader becomes effectively a co-author of the text, constructing the text collaboratively from a kit supplied by the author.

- The text printed in a codex book is fixed, whereas in hypertext the text changes every time it is read, depending on which link is activated. Therefore the meaning and experience of the hypertext varies with each reading.
- In the hypertextual environment, readers can make new connections that the original author did not anticipate.
- Readers are not passive recipients of meaning but are active meaning-makers interactively engaged – arguing, agreeing, reading against the text, and sometimes posting their own texts in response.
- Readers are not solitary and silent but are participants in a virtual community of other readers, linked by e-mail, newsgroups, electronic conferences and real-time chat.

This theorizing about the role of the reader of hypertext sounds compelling, as long as we acknowledge that most of it holds true as well for readers of printed text. Now that we can observe readers doing things in the electronic environment, it seems to be easier to identify the variety of reader-related activities performed and to recognize that they have been happening all along. In fact in the case of those genres that have migrated into a digital format without being redesigned – notably novels and monographs – readers are becoming aware, sometimes for the first time, of strategies that they have been using to negotiate codex books but are prevented from using at this 'horseless carriage' phase of digital books. Experienced readers of printed books say, for example, that when they pick up a non-fiction book that is unfamiliar to them they perform one or more of the following tests: they take into account the status of the publisher, look at the back cover to see if the book is praised by people they respect (not possible, however, with library books, whose covers are routinely stripped), examine the table of contents, look at the bibliography to discover which discourse community this new book belongs to, check the index to see which names and concepts are included, and finally sample and skim some passages of text to check writing style and quality of analysis. Once they decide to read the book, they may choose not to start with the first chapter but to begin somewhere else, even with novels. Quite often they read with a pencil at hand and make marginal notes and cross-references, sometimes carrying on a running dialogue with the author.

It is not just scholars whose relation to printed books is interactive. Researchers who have studied how children learn to read have found that literacy is deeply embedded in the social processes of family life and depends on a range of interactive activities from writing notes to sharing stories aloud. To develop a sense of themselves as confident readers and writers, children need to have the expectation that reading and writing are routine parts of everyday life. In her longitudinal case study of six families of young children entitled *Family literacy*, Denny Taylor (1983) emphasized the role of environmental print in engaging children in literacy practices from reading street signs, labels on food and clothes, advertising logos, price tags, and menus to writing telephone messages, postcards, birthday cards and grocery lists.

Everyone accepts that sharing stories with children is a crucial element in the making of readers, but ethnographic studies make it plain that 'reading to a child' consists of a lot more than an adult reading and a child listening. The child emerges as a very active partner indeed, often choosing the book to be read, initiating the reading event, asking questions, making comments on the text or pictures, and eventually becoming a reader himself or herself in a process that educators call 'chiming in', where the child begins to memorize parts of the story and can fill in gaps. Parents help build the process of meaning-making by encouraging beginning readers to draw on prior knowledge as a strategy for making sense of new texts. Taylor (1983) says that parents 'spent much of their story-reading time relating events in the stories to the everyday lives of their children' and gives the example of Andrew (three years, eight months) and his mother sharing a story in which a small boy and an elephant are lost. Andrew's mother said, 'See how people can get lost like the little ele-

phant when they don't stay near their mommies. Right?' 'Like I did',
said Andrew, on cue. 'Right', said his mother.

How do we know about reading?

Now that it is perceived as under siege in the digital age, the process
of reading, once taken for granted, has increasingly been the object
of empirical and theoretical investigation. Thus studied, reading has
turned out to a remarkably complex and variable behaviour. It is not
one single activity, but many, with different kinds of purposes, satisfactions and required skills. Two very useful and comprehensive
reviews of the research literature on reading have been written especially for a library science audience: Janice Radway (1994) and Wayne
Wiegand (1998) each provide maps of the research terrain, including
literacy studies, reader-response theory, ethnographies of reading
and the study of print culture. Reading has been studied from every
possible angle from a variety of disciplinary backgrounds including
psychology, education, sociology, literary studies and library and
information science. Each discipline has its own idea of what counts
as a worthwhile research question to ask about reading, and each has
its own privileged methods of investigation.

Probably the most radical issue dividing researchers is this: in what
model of reading do they believe? Or to put it another way, what stories do they tell about reading? Do they see readers as essentially passive objects controlled by active texts, or do they see readers as
actively involved with constructing the meaning of texts? In the so-
called 'text-active' model the text *does* something to its readers. There
is something 'in' the text itself that determines the reader's response.
The text is comprised of fixed and determinate textual features that
are undeniably *there* and have predictable effects on readers. The text
is stuffed with specific messages, beauties or effects, and the reader's
job is to extract them – the Little Jack Horner approach to reading,
as Northrop Frye once put it. Research on propaganda and pornog-
raphy is often undertaken from within this model, as is research on

mass media audiences that charts ratings and uses content analysis to figure out which messages audiences are supposed to be receiving from popular programmes (see Mosco and Kaye, 2000). Titles for this type of research often include terms like 'impacts' or 'effects', and the impacts are often thought to be bad.

Another story of reading gives a lot more power to the reader/viewer and puts emphasis on the ways in which communities of readers make sense of what they read or view. In reader-response criticism and new audience research, reading/viewing is seen as a transaction between a text and an active reader/viewer. The meaning is not so much there in the words or on the screen as it is constructed by the reader on the basis of his or her past experiences with reading texts and with living in the world (Fish, 1980). Meaning is thought to be constituted by the reader's activity in bringing certain horizons of expectations to the text, in selecting which features of the text to attend to, and in responding to these features. The balance shifts: meaning, once thought of as what the text gives, becomes something that the reader takes. If you ask different readers to read the same text, you have to take into account the fact that each reader is creating from the words on the page a different meaning. Furthermore the same reader may read the text differently on different occasions and at different stages in her life. Typically research conducted within this second model views reading as empowering, with the reader in charge, deriving benefits that range from local problem solving to the construction of an identity. In *The story species: our life-literature connection*, literary critic and family therapist Joseph Gold claims that every human being is engaged in making and remaking a life story or identity out of bits and pieces derived from experience and filtered through narrative structures. He views reading fiction as 'the best tool we have for helping ourselves develop a fully functional "I"' (Gold, 2002).

Selected research findings about reading

- Studies conducted in Canada and the USA have consistently found over periods of decades that 'heavy readers' are more likely to be female than male, to be younger rather than older, and to have achieved a higher educational level than the population at large (Book Industry Study Group, 1984; Cole and Gold, 1979; Gallup Organization, 1978; Watson et al., 1980).
- The single most striking characteristic of skilful readers is that they speed through stretches of text with apparent effortlessness (Adams, 1990).
- Boys take longer to learn to read than girls do. Once they are able to read, boys spend less time reading than girls do and are less likely to value reading for pleasure. On the other hand boys do better than girls at information retrieval and work-related literacy tasks and are far more likely than girls are to read for utilitarian purposes. Significantly more boys than girls describe themselves as 'non-readers' (Smith and Jeffrey, 2002).
- Access to books and other interesting reading materials is a critical factor in becoming a good reader. McQuillan (1998) concludes, 'There is now considerable evidence that the amount and quality of students' access to reading materials is substantively related to the amount of reading they engage in, which in turn is the most important determinant of reading achievement'. Libraries have an important role to play because low-income families especially are likely to have very few books in their home.
- On websites text attracts attention before graphics. A study using eye-tracking equipment found that readers of online news sites first looked, not at photos or graphics, but at headlines, article summaries and captions. Readers spent 22% of the time looking at images other than banner advertising, and 78% of the time on text (Poynter Institute, 2000).
- On the web users do a lot of brief scanning, foraging quickly through many article summaries, but when their interest is caught

they will dive into a particular topic or article in depth. Users engage in interlaced browsing, frequently switching among alternate sites (Poynter Institute, 2000).

- Significantly large numbers of Americans say that they use information found on the web when making important decisions related to education and job training, investments, big ticket purchases and healthcare for themselves or those close to them. Based on a survey of 1415 internet users in January 2002, the Pew internet and American Life study finds that 14 million American internet users say that the internet was crucial or important in upgrading their job skills, and 11 million say they used information found on the internet to help cope with the illness of a loved one (Pew Research Center, 2002).
- A study by JSTOR indicates that university students use online versions of journals 20 times as much as they use the corresponding paper articles bound in periodical volumes in library stacks. JSTOR (www.jstor.org) is a project that has digitized dozens of runs of scholarly journals, some having issues that go back more than 100 years (cited in Guernsey, 2000).

Conclusion

Although reading research has made great progress in recent years and has shown us that the phenomenon of reading is complex, multi-faceted and often contentious in terms of the perceived process, there is still much that remains to be studied. This chapter has highlighted some of the areas that both warrant further study and that also must be remembered in the headlong rush to electronic texts. Most importantly, we must recognize that the connection between voluntary reading and powerful literacy is that people learn to read by actually reading, not by simply scanning. But there are many kinds of reading and many kinds of literacies, each suited to a specific context or desired outcome; we must not forget this at a time when 'reading' often becomes synonymous with 'scanning' of electronic text, or when deep reading is seen as old fashioned or no longer relevant in the electronic age.

In addition, we must remember that reading in almost any context can be, perhaps should be, an intensely social activity. In the traditional print environment one may read in isolation, but the act of reading generates emotions and ideas that are shared with others both formally and informally. In an electronic environment it is both similar and different. Here, readers can make new connections that the original author did not anticipate and does not manipulate; rather, the reader is less a passive recipient of text than an active meaning-maker, participating in a range of virtual communities of other readers.

All of this, and other issues raised in this chapter, mean that we must not dismiss the act of reading as antediluvian but rather as an essential component of understanding in the electronic age. We must take this reality with us as we seek to advance our understanding of the impact and meaning of texts in all forms in the 21st century.

References

Adams, M. J. (1990) *Beginning to read: thinking and learning about print. A summary*, University of Illinois at Urbana-Champaign, Center for the Study of Reading.

Birkerts, S. (1994) *The Gutenberg elegies: the fate of reading in an electronic age*, London: Faber and Faber.

Book Industry Study Group (1984) *1983 Consumer Research Study on reading and book purchasing: focus on adults*, New York: Book Industry Study Group.

Boyarin, J. (ed.) (1992) *The ethnography of reading*, Berkeley, CA: University of California Press.

Calvino, I. (1981) *If on a winter's night a traveller*, trans. by William Weaver, Toronto: Lester and Orpen Dennys.

Chelton, M. K. (2001) When Oprah meets e-mail: virtual book clubs, *Reference and User Services Quarterly*, **41** (1) (Fall), 31–6.

Cole, J. Y. and Gold, C. S. (eds) (1979) *Reading in America: selected findings of the Book Industry Study Group's 1978 study*, Washington, DC: Library of Congress.

Darnton, R. (1982) What is the history of books?, *Daedalus* (Summer), 65–83.

Epstein, J. (2001) Reading: the digital future, *The New York Review of Books* (5 July), available at
www. nybooks.com/articles/14318
(accessed 29 May 2002)

Fish, S. (1980) *Is there a text in this class? The authority of interpretive communities*, Cambridge, MA: Harvard University Press.

Gallup Organization (1978) *Book reading and library usage: a study of habits and perceptions,* conducted for the American Library Association, Princeton, NJ: Gallup Organization.

Gold, J. (2002) *The story species: our life–literature connection,* Markham, ON: Fitzhenry and Whiteside.

Guernsey, L. (2000) The library as the latest web venture, *The New York Times*, (15 June), Sec D, 1.

Gurak, L. J. (2001) *Cyberliteracy: navigating the internet with awareness*, New Haven, CT: Yale University Press.

Healy, J. M. (1990) *Endangered minds: why our children don't think*, New York: Simon and Schuster.

Jenkins, H. (1998) *From Home[r] to the holodeck: new media and the humanities.* Keynote address presented at the Post Innocence: Narrative Textures and New Media Conference, hosted by the Trans/Forming Cultures Project at the University of Technology, Sydney, Australia, October 3, 1998, posted December 6, 1998, available at
http://media-in-transition.mit.edu/articles/australia.html
(accessed 29 May 2002)

Krashen, S. D. (1993) *The power of peading: insights from the research*, Englewood, CO: Libraries Unlimited.

Landow, G. (1992) *Hypertext: the convergence of technology and contemporary critical theory*, Baltimore, MD: Johns Hopkins University Press.

Lanham, R. (1993) *The electronic word: technology, democracy and the arts*, Chicago: University of Chicago Press.

Long, E. (1992) Textual interpretation as collective action. In Boyarin, J. (ed.) *The ethnography of reading*, Berkeley, CA: University of California Press, 180–211.

Lynch, C. (2001) The battle to define the future of the book in the digital world, *First Monday*, **6** (6) (June), available at www. firstmonday.dk/issues/issue6_6/lynch/ (accessed 29 May 2002)

McQuillan, J. (1998) *The literacy crisis: false claims, real solutions*, Portsmouth, NH: Heinemann Educational Books.

Mosco, V. and Kaye, L. (2000) Questioning the concept of audience. In Hagen, I. and Wasko, J. (eds) *Consuming audiences? Production and reception in media research*, Cresswell, NJ: Hampton Press, 30–46.

Murray, J. H. (1997) *Hamlet on the holodeck: the future of narrative in cyberspace*, Boston, MA: MIT Press.

Nunberg, G. (1993) The places of books in the age of electronic reproduction, *Representations*, **42** (Spring), 13–37.

O'Donnell, J. J. (1998) *Avatars of the word: from papyrus to cyberspace*, Cambridge, MA: Harvard University Press.

Pang, A. S.-K. (1998) Hypertext, the next generation: a review and research agenda, *First Monday*, **3** (11), available at www.firstmonday.dk/issues/issue3_11/pang/ (accessed 29 May 2002)

Pew Research Center (2002) Internet and American Life Project. January 2002, available at www.pewinternet.org/reports/ (accessed 29 May 2002)

Poynter Institute (2000) [Eye-tracking study conducted by the Poynter Institute and Stanford University], May 2000, available at www.poynter.org/eyetrack2000/ (accessed 29 May 2002)

Radway, J. (1994) Beyond Mary Bailey and old maid librarians:

reimagining readers and rethinking reading, *Journal of Education for Library and Information Science*, **35** (4) (Fall), 275–96.

Ross, C. S. (1999) Finding without seeking: the information encounter in the context of reading for pleasure, *Information Processing & Management*, **35**, 783-799.

Ross, C. S. (2001) What we know from readers about the experience of reading. In Shearer, K. D. and Burgin, R. (eds), *The readers' advisor's companion*, Englewood, CO: Libraries Unlimited, 77–95.

Schwartz, L. S. (1996) *Ruined by reading: a life in books*, Boston, MA: Beacon Press.

Smith, M. and Jeffrey, D. (2002) *'Reading don't fix no Chevys': literacy in the lives of young men*, Portsmouth, NH: Heinemann Educational Books.

Snyder, I. (n.d.) Hyperfiction: its possibilities in English, available at www.schools.ash.org.au/litweb/ilana.html (accessed 29 May 2002)

Taylor, D. (1983) *Family literacy: young children learning to read and write*, Exeter, NH: Heinemann Educational Books.

Watson, K. F. et al. (1980) *Leisure reading habits: a survey of the leisure reading habits of Canadian adults with some international comparisons*, Ottawa: Infoscan.

Wiegand, W. A. (1998) Introduction: theoretical foundations for analyzing print culture as agency and practice in a diverse modern America. In Danky, J. P. and Wiegand, W. A. (eds), *Print culture in a diverse America*, Urbana, IL: University of Illinois Press, 1–13.

6

Judging a book by its cover: e-books, digitization and print on demand

Shirley Hyatt

Introduction

> When I look at the book, the image I see keeps changing. It is a bundle of papyrus leaves, a roll of parchment, a stitched set of pages, and a flow of images on a screen. Perhaps, like beauty, 'the book' exists only in the eye of the beholder – that is, each of us has a subjective notion of its essence, even though it is constantly passing through mutations.
>
> (Darnton and Kato, 2001)

Seen from the perspective of 2002, it is clear that 1999 and 2000 were 'binge years' for electronic books and electronic publishing: news reporters proclaimed the goings on of authors, of publishers, of book companies; authors experimented with distribution channels and software tools; publishers launched e-book initiatives; and librarians evaluated, discussed and loaned e-books. Books had never before been so telegenic; new initiatives sprang up everywhere. Compared with the excitement of those years, 2001 and 2002 reflect an emerging conservatism, concern and cynicism. Among the e-conservatives there is smugness; among the e-liberals, quiet. Some initiatives have been withdrawn; some authors have declared retirement, and several businesses are no longer with us. Nevertheless, we sit in 2002 with sure knowledge (confidence on the part of some, resignation on the part of others) that the next five years will bring great strides in e-bookmanship. New business rules, new publishing expectations and new audience expectations generated by a digital world are transforming the book world steadily and surely. Accordingly, this chapter examines some of the forces that are working to transform the book world that are affecting readers, authors, publishers, distributors, sellers and collection managers, and that will, ultimately, move books squarely into an electronic world.

Several components make up the electronic reading experience, so a few definitions are in order. The first component is hardware: one reads an e-book on a desktop computer, a laptop, or some smaller wireless device – tablets, personal digital assistants such as Palm and Handspring, handheld PCs, and dedicated reading devices such as the RCA Gemstar REB. The reading device may even be a learning toy such as the Leap Pad dictionaries found in toy stores – all referred to in this chapter as 'reading appliances'.

The second component is software that facilitates the searching, navigation, font appearance, functionality and presentation of information. Most e-book companies require the use of proprietary client-side reading software to view and manage their books independently of the web. A few, like netLibrary, require standard browsing software to view centrally located e-books. In this chapter both types are referred to as 'reading software'.

'E-book' refers to content that has been made available digitally and electronically. This includes text that has been converted from print to digital form, digital images, digital audio books and content that has been created from the 'get-go' in a digital mode. Their formats, the software used to create the e-books, and languages in which they are coded vary considerably, and the term 'e-book' encompasses all of these. E-books may be distributed via the web, by CD or floppy disk, via an aggregator dedicated to e-books, by a publisher or author site, by not-for-profit digitization projects, via software-oriented download sites, or by a wireless connection from a mobile device.

This chapter is about e-books themselves – not 'gizmos' or reading software. The discussion in this chapter focuses on descendants of the traditional book industry, evolutionary offspring of that entity that keeps changing, that constantly passes through mutations, and that might possibly just be something that exists only in the eyes of the beholder (Darnton, 1999). Fundamental to this discussion is a belief that the e-book business – including authoring, publishing, library management, and reading – is not significantly different from the movement toward digitization that has taken place over the past 30 years.

Trends in authoring, publishing and bookselling

By all accounts we have reached a crisis in scholarly publishing. While faculty at research institutions 'donate' their articles to journals as a prerequisite to managing their careers, their institutions and libraries are required to buy back the work, in the form of journal subscriptions, at exorbitant rates. The exorbitant increase in the price of periodicals has forced libraries to cut back on their purchases of monographs. The cuts in library budgets for monographs has made university presses reluctant to publish books in fields where demand is weakest.

> It used to be the case that university presses (and other small presses) could count on selling 800 copies of a monograph, and could count on libraries to ensure that their books would not be published at a loss. Today the figure is 400, often less, and not enough in any case to cover costs. These publishers can no longer be sure of selling books that would have been irresistible to librarians twenty years ago.
>
> (Darnton and Kato, 2001)

As a result, some of the finest young scholars are finding it impossible to publish their research, though ground rules for employment ('publish or perish') remain unchanged. Eventually, it is surmised by some, the affected disciplines will atrophy from lack of talent able to prove itself in the traditional ways.

The crisis is equally telling in trade publishing. Book profits traditionally have been based on a serendipitous combination of best-sellers and front lists (titles issued within the past year), and mid- and back lists (that is, trade books with artistic and intellectual aspirations that were not earmarked in advance as blockbusters – principally literary fiction and serious non-fiction). Mainstream book publishers are finding it especially hard to make a profit these days, hard to publish and cultivate back-list authors, hard to maintain quality mid- and back-title lists. There are four principal indicators of the seriousness of this problem.

First, almost all of these publishers have been purchased by huge entertainment media conglomerates, who are suspected of inappropriately treating books as if they were a commodity no different from their other products. Some accuse these conglomerates of being too greedy, intentionally focusing on books that will be huge block-busters, while ignoring cultivation of mid-list authors. Some quality imprints have been discontinued by these corporations, and author contracts cancelled. In 1997, for example, HarperCollins, a unit of Rupert Murdoch's News Corporation, announced plans to fold its Basic Books imprint into its main adult trade list; then in the following June HarperCollins cancelled the book contracts of more than 100 authors. Nevertheless, studies show that these conglomerates, including HarperCollins, are producing just as many mid-list titles, yet are actually profiting less per title than in the glory days of publishing (Kirkpatrick, 2000).

Second, in a tradition dating back to the American Depression intended to protect booksellers from the risk associated with offering titles that they are not able to review in advance, publishers permit full refunds for titles that do not sell in bookstores within 90 days.

Books that do not sell within 90 days are at risk of being taken off the shelves, and returns by chains have reached record proportions, spiking to 37% in 1997. This affects the profitability of books.

Third, in the USA inventories are taxed, creating a disincentive for publishers to retain stock. Books have become less of an asset to be stockpiled and more of a fluid commodity to be moved as quickly as possible.

Finally, with over 16,000 public libraries in the USA alone, the library market could shore up well-reviewed books that would otherwise have borderline profitability. Now, with the decline in library monograph budgets, these titles are much less likely to generate profit for publishers.

Entangled with the changes in publishing are the emergence of huge mega-bookshops and the demise of small, independent bookshops. Independent bookshops traditionally have a strong role in hand-marketing and selling mid-list books to the public. Independents complain, with justification, that they are being driven out of business by the large booksellers, to whom publishers have given favourable discounts. Indeed, independent booksellers in the USA reduced from 5100 in 1993 to 3500 in 2000 (Kirkpatrick, 2000). The American Booksellers Association, the independent store owners' trade group, successfully brought anti-trust suits against six of the largest book publishers for unfair competitive practices.

The mega-bookstores, with their significant negotiating power, levy payments from publishers for favourable merchandising. Displaying books on front tables, for example, may cost publishers US$10,000 per title per store. The publishers in turn select only blockbuster books for this exorbitantly priced merchandising treatment, leaving fewer marketing dollars for mid-list books. Chainstore merchandising policies help turn consumers' attention away from mid-list titles and toward those elite books that are backed by heavy marketing budgets.

At the heart of the trade publishing crisis is a change in book merchandizing. Book publishers are producing as many mid-list books as in the past, and the new title count continues to grow. However, the mid-list market share is decreasing, and its share of merchandizing dollar is dwindling.

> Mid-list book sales have declined as a percentage of total sales and, over the last ten years, in absolute numbers as well. Chains, superstores, and Internet booksellers have made an enormous range of mid-list titles available, but without heavy marketing support, these books get lost. Attention in the chains and superstores often comes at a high price, and publishers are willing to pay it usually only in the case of obviously commercial books. (Kirkpatrick, 2000)

The problem for authors is less one of getting a publishing contract than it is one of getting the publisher's marketing dollars, which are necessary for being reviewed, put on retail shelves, purchased, read, and retained in print. Independent booksellers' voices have become fainter; and the dwindling of their numbers and share of book sales has hurt mid-list books, as has the dwindling library monograph budget.

Consolidation of journal publishers, consolidation of book publishers, consolidation of booksellers, and consolidation of book merchandising into a small subset of bestselling titles, result in a myriad of interdependent issues. These changes and the impact they have on books – the profitability of mid-list publishing, the cost-effectiveness of keeping titles in print, the need for alternative publishing channels, the need for alternative, cheaper, marketing channels for books – form the backdrop for a booming interest in e-books. These issues all have solutions that reside in the digital file and so drive the exploitation of that file.

In contrast to the days of print only, today there is a dazzling diversity in channels for information distribution. *The Wall Street Journal*, for example, produces its paper version (with local editions) and also a web version, a Palm version, an e-book version, and an Avant-Go (also known as web clipping service) version.

Concurrent advances in technological underpinnings of publishing, selling and merchandizing bring opportunities for the exploration and exploitation of technology. There has been much talk of authors and researchers taking control of the publishing process. In scholarly publishing this has come to be known as the Open Access Initiative or the Free Online Scholarship movement, and its aim is to make peer-reviewed research articles in all academic fields freely available on the internet.

In the book world the trend – if it can be called that – is towards 'self-publishing', wherein book authors either edit their own work or contract for their own production, and make their own books available using the web as an essential marketing and distribution channel. More often than not, the book is an e-book. Stephen King's *Riding the bullet* and *The plant* experiments jump-started awareness of this e-book authoring movement. Another striking example of self-publishing is Seth Godin's *Unleashing the ideavirus*. Godin wrote the book and made it freely available on the web. After a gestation period on the web, he made printed hardbound copies available. With web promotion but no advance capital, and with 28,000 pre-publication orders, his book was in profit within nine weeks of the print version's release (Abbott, 2000).

Dot.com companies have been launched to support this trend, and they offer a variety of services at differing rates; some mainstream publishers have launched self-help services for authors as well. Since the mid-list authors suffer considerably from lack of merchandising, this gives them a significant opportunity to market themselves if they have the energy, time and resources to do so, and without significant cash outlay. Many publishers, acknowledging the scarcity of merchandising funds, at the very least encourage authors under contract to market their own books as much as possible.

We already have a significant amount of cross-merchandising; advertisements from bookstores are, for example, substantially funded by publishers. With publishing now owned by media giants, there are new opportunities for cross-promotion and e-commerce.

The conglomerate publishers envisage links between their TV and film properties and their book content sites. Imagine watching 'Sex and the City', and in one corner an AOL logo comes on: 'Buy the book, click here' (Hilts, 2000). These will be convenience buys for consumers, and libraries may well want to find ways to engage in parallel opportunities.

Trends in printing, finishing and distribution

One cannot underestimate the importance of the ubiquity of laser printing, colour printing, digital printing and networking in today's environment. The number of printing machines worldwide, now at 10,000, is expected to double every two years at the present rate of growth (Kenji, 2001). Commercial quick-copy centres are a staple on every campus. Desktop computer owners routinely own desktop printers, and the quality, speed and prices of these have improved dramatically. Individual consumers have been printing on demand for years under US fair use guidelines and perhaps also illegally. Academic libraries have noticed steep drops in library printing recently, doubtless as a result of consumers printing within their own homes and offices.

There are several recent trade innovations as well: improvements in quick binding, and fast finishing. These technological improvements offer the potential for new distribution methods and new business models. A digitally printed page in 1988 cost 29 cents; it now costs 10 cents or less.

In the past publishers estimated the number of copies needed, then printed that number, then shipped them to distributors and then from there the copies were shipped to booksellers. The 'print and distribute' model results in significant costs for the shipping, warehousing and reshipping of books, and then the return shipping of unsold books. Networking combined with high-speed, high-quality printing and binding/finishing enables the publication industry to switch to a 'distribute then print' model, obviating the need for print overruns and warehousing of stock. It also obviates the need for shipping, since printing centres around the world can be sent digital files. Print on demand, as this is called, is the ability to use a digital database and actually print from it a physical book that looks like and feels like a normal physical book.

The goal concept is that a person will be able to go into a bookshop, to a kiosk or a website, decide that he or she wants a book and have it printed on demand – in a matter of minutes. Today the book would have to come from a central warehouse, but in the next couple of years, with distributed computer processing technology developing as it has been developing, we will be able to print that book in a shop (or a library) and have it bound while drinking coffee. This is an 'inventory on demand' model. The independent presses have jointly negotiated purchase of some print on demand systems for their members. Barnes and Noble anticipates having print on demand capacity in its stores in a few years – but they will have manufacturing capacity for music on demand sooner.

Some publishers have discussed developing a huge central catalogue of digitized books, complete with the metadata needed to search for and identify works, facilitate print on demand worldwide, and (presumably) manage rights and administrative record-keeping for this shared, co-operative, print-on-demand utility (Epstein, 2001). This catalogue would interact with point-of-sale print-on-demand machines.

Print on demand, though it is often thought of as a single, monolithic entity, is actually many different kinds of machines and systems being applied in different ways to different publishing markets: educational and technical publishing, libraries, computer manuals, and now reaching into the general trade (Hilts, 1998). In the educational market it permits last-minute changes to textbooks to ensure that text is up to date with important discoveries and events. It also enables local versions where, for example, particular social issues may be viewed in a unique way. It permits instructors to create courseware customized for their syllabi and produced in paper form. In libraries it has potential to reduce the number of acquisitions the library buys;

fewer books need to be purchased prior to user requests, since the book can be made available in short order. In commercial and not-for-profit publishing it has potential to keep back-list titles available for new requests; a book need never go out of print. With the resolution of digital rights, print on demand may also have potential for the creation of customized books.

It can also enable printing books far from their original source. A book that can be transmitted over the web, and printed locally, can evade export and customs fees, and can also evade customs censorship. This is a critical asset in countries where cultural heritage may otherwise be lost and where cultural policing is commonplace. The Kurdish language, for example, is forbidden in Turkey. With print on demand, a Kurdish–Turkish dictionary has been created elsewhere and made available in Turkey – the print version, shall we say, of Radio Free Europe (Curman, 2001). The 1979 Iranian revolution was in large part fomented by smuggling audio tapes into Iran made by the Ayatollah Khomeini while exiled in France. A single smuggled tape could be copied and heard by millions. Today, other ayatollahs and politicians use digital technology to challenge the very state brought about by Khomeini's revolution (Mahloujian, 2001). Words will always be a powerful force, and networking and print on demand expand their reach.

To summarize, printing on demand:

- offers a 'distribute then print' model to reduce shipping and warehousing overheads
- enables jumping borders to subvert censorship
- facilitates short print runs, which salvages the market for low-demand genres such as poetry, literatures of minority cultures and endangered languages, and print runs adapted to a highly specific and targeted market
- offers liberation of literature from the bestseller mentality
- facilitates personalized and customized printing, which is important among other features for accessibility

- enables an items to be 'always in print', or at least in print as long as the digital format is supported
- provides competitive benefit to independent presses.

Shored up by an available web-based catalogue of titles, the independent presses and mega-bookstores do not need a large book stock inventory. Print on demand also can contribute to compiling editions in a new way, reviving out-of-print titles, achieving cultural policy goals, restoring the vibrancy of small bookshops, meeting the needs of foreign-language readers, and bringing literature into the world at large.

The changing nature of book production and distribution is just another step in a long history of change in the book world. The book as an object has undergone innumerable metamorphoses, and so has our notion of the book. During the 16th century stationer-booksellers typically maintained workshops in which binding was carried out. In the next century these merchants served as brokers for bindery services, and the actual binding was done by independent master binders. Note that it was not the publisher who bound the books. Book buyers purchased loose bunches of printed paper (folios and signatures), and booksellers served as middlemen for binding. Indeed, for significant periods of history booksellers were legally enjoined from selling bound books directly to the public unless registered as a stationer. Purchasers could select common bindings – trade bindings, as they came to be called – or more sumptuous bindings such as silk and damask. The book collector chose bindings to suit his/her taste and income, or, as some did in the 17th century, to ensure that all their fine books looked aesthetically identical on the bookshelf.

Only in the 19th century, with the industrial revolution, did this practice change.

> Although cloth binding as we know it was first adapted to bookbinding in 1823, 'a style of binding uniform for all copies of the same book' did not appear until around 1830, when

machinery was introduced to letter the clothbound cases that could be fitted over the printed guts of a book. This development ushered in a new chapter in the way books were made and sold. Whereas the bookseller would bind or have bound, by hand of course, only as many copies as were likely to be sold in the immediate future – a form of just-in-time manufacturing – with the advent of machinery the publisher itself began to bind an entire edition of a book in the common style of the time.   (Petroski, 1999)

Books, in other words, were not entities defined by the publisher as a specified bound unit linked to a title. A bound edition was a function of the owner's taste and income – and interior decor – and not the publisher's predetermined construction. One cannot help but ponder the sameness of this business model and the just-in-time binding model we now face with electronic books.

Blurring of boundaries

With the rapid changes brought about by the digital file has come a blurring of boundaries in the division of labour that developed over centuries of technological advances related to the printed book. Authors are experimenting with being their own publishers. Entertainment and media companies are exploring new roles as book publishers. University presses are assuming titles that are increasingly under-served by trade publishers as a way of leveraging their core competencies and so becoming mid-list publishers. With print on demand, publishers are becoming printers, and printers are becoming publishers and binders. Publishers are also becoming direct booksellers via the web. Libraries are joining the fray, and as they digitize items, collect and make widely known local collections via the web, they assume the publisher's role. Readers themselves, by editing and compiling works to suit their own individual needs, are becoming authors of new, singular, works and of anthologies. Authentic books

– books produced with the explicit approval and quality check of a publisher with rights to the item – blur into pirated print copies. Authorized copies of the electronic work may blend into unauthorized versions. Text, images, audio formats are increasingly synchronized or integrated, and books simply cease to have any boundaries whatsoever.

It is likely that all the specialists will eventually return to their core competencies, but armed with better tools. Electronic books, on the other hand, are here to stay. Networked distribution of an electronic document, with or without print output, has already become a staple of the information seeker's portfolio.

Despite the brouhaha of recent years about e-books, they have already made strong inroads into the information sector: it is commonplace for electronic abstracting and indexing databases to have replaced paper reference works in libraries, businesses, government and private homes. Hard-copy law books have all but disappeared; people in law firms do not use them any longer, relying on electronic books or online services almost exclusively. Online and CD-based encyclopedias have replaced or augmented bound encyclopedias. Desktop software, previously distributed via CD-ROM with accompanying paper-based technical documentation, is today almost exclusively distributed with electronic documentation. CDs have made inroads into many formerly paper-based book segments such as nature guides, tutorial products and self-learning. E-journals have made huge 'e-strides' in the past few years, and studies show that library users greatly prefer them. Computerized 'help systems' have supplanted external paper-based reference works. Websites of all sorts have replaced paper-based communications ranging from wedding and holiday albums to technical documentation. Apparently, as the paper and bound form of communication is eclipsed by an electronic version, the successor ceases to be thought of specifically as an e-book.

Trends in content, reading and e-books

Content, as ever, remains the sine qua non of the book. Ultimately a book is nothing without its content, and the content is nothing without some means to enable the human brain to interact with it. The electronic form of the book is simply one more modality that enables a person's brain and the content to interact.

When the codex began to replace the scroll, paper had to be improved, and binding processes honed. Basic organizational tools of pages, sections, chapter, and volumes, and embedded metadata such as title pages, had to be invented by authors and adapted over time. Never before could the reader flip through an entire work; never before did the reader have free hands with which to write; never before was navigation so enhanced as it was by tables of contents, indexes and page numbers. Storage containers had to be developed. All of this took time, invention, promotion and user acceptance (Chartier, 2000).

Several authors have discussed the adaptive change over many generations from 'intensive reading' to 'extensive reading' (Darnton and Kato, 2001). When readers owned only a few books, they read them over and over. When all sorts of printed material became available, readers changed their activity pattern, reading a text once and then moving on to the next one. Undoubtedly we are in the process of a parallel shift illustrated by our hopping from one website to another. Surfing is thought to be a right-brained activity, whereas reading words on a page is traditionally thought to be a left-brained activity.

The use and manipulation of books is not just a matter of immersion, but of scanning, clipping, bookmarking, forgetting, later reference and 'manhandling'. One study about how readers interact with texts has found that

> scholars read relatively little of most of the books that they review for their work . . . . The introduction, a few pages to a couple [of] chapters, and the bibliography, footnotes, and index are browsed or read in some other books . . . . Only a

few books are read at great length . . . . They want to be able
to flip between pages, to follow the line of reasoning, to move
from reference to footnote, from index to text easily . . . . They
feel that given navigational flexibility, speed, and design that
takes advantage of interactivity, as well as substantial collections, scholars would increase their use of online books.

(Summerfield, 1999, cited in Hughes and Buchanan, 2001)

The irony is that, with all the transformation in word processing, production, digitization, distribution and presentation of information,
books today are not all that different from 40 years ago, or for that
matter 1500 years ago when the codex first emerged. Though digitization brings with it unique capabilities of separating the work from
the object that supports its transmission, book text continues to be
pretty much the same sequential material it has always been, with a
few illustrations interrupting the flow of text. The electronic file
remains a snapshot of the pages, sections, chapters and volumes of
which the printed book consists. We are simply pouring post-print
text into a digital container without restructuring it and without
rethinking the medium for the powerful message the medium can be.

The possibilities of the electronic text should increasingly inspire us
to organize it differently. An example of the possibilities is the elegant
and intriguing electronic book *Midnight play* by Pacovská (1999), which
translates a classic children's picture book into an interactive, kinetic,
multi-sensory experience. A new wave of content can emerge involving new hypertext authoring, multimedia authoring, and hypertext
teaching mechanisms. The hyperlink permits us to make non-linear
relationships among sounds, text and imagery, and among an unlimited number of documents so organized. It is possible that writers and
curriculum designers will 'atomize' or deconstruct the content and
restructure it into personalization mechanisms to permit readers and
learners to follow just that path that their personal cognitive style
requires. Scholarly works can be layered, with a concise report on the
top layer, followed by a second layer containing 'expanded versions of

different aspects of the argument . . . as self-contained units that feed the topmost story. The third layer would be composed of documentation . . . each set off by interpretive essays'. Subsequent layers 'could be pedagogical, consisting of suggestions for classroom discussion and a model syllabus . . .', and so on (Darnton, 1999).

Books are already being developed that rely heavily on the internet for added content; books that dynamically update themselves can also emerge. The accompanying danger, of course, is the spectre of e-books dynamically or surreptitiously censoring themselves and each other. Also, with the electronic text, intertextuality – interweaving of various texts – is easy. A unique rapport can develop between the author of the document and the reader: the author establishes the basic elements with multiple pathways and relationships, and the reader engages in a creative, cognitive ordering of elements. One might say that electronic writing adapts to our needs, while print adapts to the needs of mass production.

There may well come a day when the population looks back on the quaint period when scholarly documentation about dances lacked holographic references to the actual event, when pronouncing dictionaries did not pronounce, when readers had to purchase audio books separately from paper books, when one interacted with texts through keystrokes and writing rather than through machine-audible verbal instructions, when patrons viewed metadata separate from the item represented by the metadata, when books were not able to be tailored to the lighting, eyesight and dexterity of the reader, and when books did not afford readers a multitude of choices, interactions and opportunities to discuss, query and create.

In the literature of the e-book, there seem to be three main benefit strands that motivate the use of electronic documents. The first is convenience. Readers of trade paperbacks who also carry PDAs, for example, love the convenience of always having reading materials on hand wherever they are, for unforeseen reading opportunities. The second is the improved access and search/navigational abilities that the electronic document brings with it, including immediate gratification via web distribution, full-text searching, and inter-relatedness with other reference materials such as dictionaries.

Third – mostly projected at this time – are the ways that the electronic books defeat the disadvantages of physical books. Loved as they are, physical books are far from ergonomically and aesthetically perfect: people with vision problems, fine motor co-ordination problems, and wrist strength problems have significant reading challenges with physical books; people who cannot bend down or reach high, or who are wheelchair bound have significant access problems with book stacks. Surprisingly large areas of the USA, and indeed of other countries, are not served by libraries, and many parts of the world are not presently served by bookstores or news stands. Only approximately 10% of texts published today are ever made available in Braille or as Talking Books, and they are only available some time after the book is first published for the sighted reader. Furthermore, Braille on paper and Talking Books on audio cassette cannot be used as flexibly as print or digital books. In short, neither paper, nor codex, nor book stack, nor traditional library are flexible and universal; none promote the vision of 'books everywhere, for everyone, at every time'. Electronic books and digital libraries, however, can hold that promise, and this is certainly a vision worth working toward.

Conclusion

Most currently created literature, scholarly or otherwise, resides in electronic form at various phases of its production. These digital files are exploited because various partners in the chain – from authors to publishers to distributors to collection managers to readers – need the benefits and features that are available in the electronic medium. Benefits include reduction of printing and distribution overhead, merchandizing advantages, print-on-demand opportunities, playfulness of the medium, and expansion of cognitive opportunities.

Electronic books have already made strong inroads into our lives. Digitized text has been with us more than 30 years. We overlook this, and we lose sight of 'the book' when books transform themselves from paper to electronic form and as they become transparently integrated into our daily activities. Of course, we have not reached the future as yet. We still lack the regular convergence of those additional ingredients necessary to integrate the e-book into our lives with even greater regularity and consistency: effective, efficient. human-oriented search software that will enable us to locate texts quickly and with ease when we do not particularly know what we are looking for; website designs that enable us to find what we are looking for quickly, when we do know what we are looking for; portals that are reliable and well maintained; decision support; improved metadata; unique identifiers to distinguish one object from another; credibility mechanisms, to ensure that a retrieved text can be trusted and is from an authoritative source – to name just a few matters awaiting future developments.

Publishers will need a shared sense of the standards (formal or de facto) for e-book formats, and those standards will need to be compatible with the reading appliances and reading approaches preferred by readers. Rights need to be sorted out, agreements will be needed, and acceptable licensing and pricing mechanisms will eventually be determined. Much of that convergence is likely to come, over time, with patience and work. Nevertheless, the book itself will likely just keep changing, passing through additional mutations, exacting further changes from us.

References

Abbott, C. (2000) Godin moves from e-book to hardcover p-book, *Publishers Weekly*, **247** (38) (18 September), 21.

Brown, G. J. (2001) Beyond print: reading digitally, *Library Hi Tech*, **19** (4), 390–9.

Burk, R. (2001) E-book devices and the marketplace: in search of customers, *Library Hi Tech*, **19** (4), 325–31.

Chartier, R. (2000) Death of the reader?, *100-day dialogue: what has happened to reading*, available at

www.honco.net/100day/02/2000-0531-chartier.html
(accessed 13 March 2002)

Connaway, L. S. (2001) Bringing electronic books (e-books) into the
digital library. In *National Online 2001: proceedings*, Medford, NJ:
Information Today, 115–20.

Connaway, L. S. (2001) A web-based electronic book (e-book) library:
the netLibrary model, *Library Hi Tech*, **19** (4), 340–9

Coyle, K. (2001) Stakeholders and standards in the e-book ecology:
or, it's the economics, stupid!, *Library Hi Tech*, **19** (4), 314–24.

Curman, P. (2001) Distance publishing: a world of possibilities, *The
Book and the Computer*, (27 April), available at
www.honco.net/100day/03/2001-0427-curman.html
(accessed 13 March 2002)

Darnton, R. (1999) The new age of the book, *The New York Review of
Books*, (18 March), 5–7.

Darnton, R. and Kato, K. (2001) The bookless future: an online
exchange between Robert Darnton and Keijo Kato, *The Book and
the Computer*, (7 June), available at
www.honco.net/100day/03/2001-0607-dk1.html
(accessed 13 March 2002)

Dorner, D. G. (2000) Blurring of boundaries: digital information and
its impact on collection management. In Gorman, G. E. (ed.),
*International yearbook of library and information management
2000/2001: collection management*, London: Library Association
Publishing, 15–44.

Epstein, J. (2001) *Book business: publishing past, present, and future*, New
York: W. W. Norton & Company.

Epstein, J. (2001) The next golden age of publishing: an interview
with Jason Epstein, *The Book and the Computer*, (23 April), available
at
www.honco.net/100day/03/2001-0423-jasonepstein.html
(accessed 13 March 2002)

Fischer, R. and Lugg, R. (2001) E-book basics, *Collection Building*, **20**
(3), 119–22.

Forrester Research (2000) *Books unbound*, available at
www.forrester.com
(accessed 13 March 2002)

Hawkins, D. T. (2000) Electronic books: a major publishing revolution. Part 1: General considerations and issues, *Online*, **24** (July/August), 14–28.

Hawkins, D. T. (2000) Electronic books: a major publishing revolution. Part 2: The marketplace, *Online*, **25** (September/October), 18–36.

Hilts, P. (1998) Approaching the point of no returns, *Publishers Weekly*, **245** (25) (22 June), 64–5.

Hilts, P. (1998) On beyond on demand, *Publishers Weekly*, **245** (43) (26 October), 36–7.

Hilts, P. (2000) Looking at the e-book market, *Publishers Weekly*, **247** (47) (20 November), 35–6.

Hilts, P. and Lichtenberg, J. (1998) Redefining distribution, *Publishers Weekly*, **245** (51) (21 December), 23–4.

Hughes, C. A. and Buchanan, N. L. (2001) Use of electronic monographs in the humanities and social sciences, *Library Hi Tech*, **19** (4), 368–75.

Jensen, M. (2001) Academic press gives away its secret of success, *The Chronicle of Higher Education*, (14 September), B24, available at http://chronicle.com/weekly/v48/i03/03b02401.htm (accessed 13 March 2002)

Johnston, D. and deBronkart, D. (2002) Digital print is coming of age, *Printing and Converting Decisions International*, 39–41, available at www.podi.org/library/pdf/Printing_and_Converting_Decisions.pdf (accessed 13 March 2002)

Kenji, M. (2001) *The book & the computer online journal*, available at www.honco.net

Kirkpatrick, D. D. (2000) *Report to the Authors Guild Midlist Books Study Committee*, available at www.authorsguild.org/prmidlist.html

(accessed 13 March 2002)

Lynch, C. (2001) The battle to define the future of the book in the digital world, *First Monday*, **6** (6), 1–49, available at www.firstmonday.dk/issues/issue6_6/lynch/ (accessed 13 March 2002)

Mahloujian, A. (2001) A vital role for electronic publishing, *The Book and the Computer*, (3 August), available at www.honco.net/100day/03/2001-0803-azar.html (accessed 13 March 2002)

Pacovská, K. (1999) *Midnight play: a fantasy theatre for young minds*, New York: Simon & Schuster Interactive.

Petroski, H. (1999) *The book on the bookshelf*, New York: Alfred A. Knopf.

Smith, T. (2001) *E-book evolution*, Public Broadcasting Service, available at www.pbs.org/newshour/media/ebooks/index.html (accessed 13 March 2002)

Snowhill, L. (2001) E-books and their future in academic libraries, *D-Lib Magazine*, **7** (7/8), available at www.dlib.org/dlib/july01/snowhill/07snowhill.html (accessed 13 March 2002)

Sperber, D. (2001) Reading without writing, Text-e.org, available at www.text-e.org/conf/index.cfm?fa=texte&confText_ID=12 (accessed 13 March 2002)

Summerfield, M. (1999) Online books evaluation project, available at www.columbia.edu/cu/libraries/digital/olbdocs/focus_spring99.html (accessed 13 March 2002)

Vanilla, J. (1993) Print . . . on demand', *Communications World*, **10** (8), 34–6.

Wilson, R. (2001) Evolution of portable electronic books, *Ariadne*, (2 October), available at www.ariadne.ac.uk/issue29/wilson/intro.html (accessed 13 March 2002)

7

The digital library and younger users

Denice Adkins

Introduction

Library literature distinguishes between two types of younger users: children, from birth to age 11, and young adults, age 12 to 18. Libraries that provide services to these groups act as an extension of the education system. *Information power* states, 'the mission of the [school] library media program is to ensure that students and staff are effective users of ideas and information . . . by providing intellectual and physical access to materials in all formats, by providing instruction to foster competence and stimulate interest in reading, viewing, and using information and ideas' (American Association of School Librarians, 1988). Indeed, the goal of any library is that its patrons understand how to use the information and entertainment resources of that library to enhance their lives.

In *Digital libraries* William Arms (2000) discusses some benefits of the digital library format over traditional libraries. The networked aspect of a digital library allows it to be accessed at the user's location and at the user's convenience, while providing access to multiple users around the world. The electronic aspect of the digital library allows information to be updated easily, indexed efficiently, and searched effectively. Artefacts stored in digital libraries are not limited to text, but can take on a variety of formats and cater to a variety of learning styles. Additionally, the digital nature of the digital library may influence the creation of information, opening new avenues for dissemination. Arms writes, 'the fundamental reason for building digital libraries is a belief that they will provide better delivery of information than was possible in the past'. Digital libraries have incredible potential to improve education and provide information to younger users. However, developers have traditionally built digital libraries for educated adults. Access by younger users is limited by economic, intellectual and linguistic barriers, and digital delivery is only one facet of the younger user's information-seeking experience.

Traditional libraries and younger users

Historically, several services have been provided by traditional libraries to children and young adults, as well as their care givers. Age-appropriate reference and information services, reading guidance, sources of informational and recreational reading, media such as musical recordings, videos, and books on tape, provision of information technology such as computers and internet access, programmes and areas for socialization are provided by traditional libraries to younger users (Willett, 1995, p.118). Libraries encourage care givers, parents, teachers and guardians to participate in literacy- and development-oriented experiences such as family reading groups. They work with schools, medical facilities and childcare organizations to bring literacy to children who do not visit the library (Ray, 1979). In providing services to younger users, traditional libraries have historically appealed both to education and entertainment values. They provide materials to their younger users both for leisure and research. They offer access to a broad range of information, written at levels that are developmentally appropriate for children and youth. Entertaining library programmes usually have some educational or acculturative value – if children are not actively learning information, they are passively learning group behaviour skills.

Five conditions for youth services librarianship have been suggested: '(1) specialized collections, (2) specialized space, (3) specialized personnel, and (4) specialized programs/services designed for youth, (5) all existing within a network of other youth services organizations and agencies' (Thomas, 1982, as cited in Jenkins, 2000). The history of library services to children and young adults is relatively recent. 'The model for youth services librarianship as it is currently configured took shape and matured in the United States' (Jenkins, 2000), beginning in 1803 with the founding of a library for youth in Connecticut. The UK enacted legislation for public libraries in 1850, and the provision of children's materials began in the late 1800s and 'became widespread during the 1920s and 1930s' (Ray, 1978). Many other developed nations started providing library services to children in the 1900s. The 'Anglo-Saxon concept of the public library' was introduced in France in 1910, and the first children's library in 1924 (Patte, 1978). The first reading room for children appeared in Iceland in 1924 (Hannesdóttir, 1978). In 1920 the Danish government enacted the Danish Public Library Act. Fifty years later, Danish libraries reported providing books, records and tapes, games, educational toys, storytimes, cultural programmes and professionally trained children's librarians (Bredsdorff, 1978). Australia began building and funding public libraries in the 1930s (Hume, 1978). Singapore established its free public library in 1958, providing separate children's and young adult areas, with books and periodicals for children, service by multilingual staff, and entertaining programmes. The children's services department worked with schools and the Ministry of Education to reach more children (Klass and Perumbulavil, 1978).

The history of library service to children in less-developed countries is not as consistent. 'People in poor, rural communities often have very little access to libraries, yet these are the areas where the majority of the socio-economic poor actually live' (Elkin, 1999). Brazilian library development for children started in 1935 with the Monteiro Lobato Children's Library. By 1978 Brazil was reported to have 54 children's libraries, 23 of which were in the capital city of Sao Paolo (Onaga, 1978). Hungarian libraries were reported as being 'eager to develop their services for children' since 1950, but provided services to only 50% of their child population in 1978 (Vargha, 1978). Nigeria's public library system dates from the 1920s; however, 'government [placed] greater priority on food, shelter, health, and education' than on library funding (Omolayole, 1978). The Tanzania Library Service was initiated in 1963. Because of limited book production in the country, books for children had to be imported from other countries; and in 1978 no trained children's librarian was available (Makafu, 1978). A recent study of South African public children's librarians found that many were not trained in children's services, while the field of school librarians had suffered drastic cuts (Hart, 2000). Public libraries interacted with local schools, but many teachers were afraid to send their classes to the library for fear of violence (Hart, 2000). These place-bound traditional libraries require that users come to them for service. The advent of the digital library raises hopes that information can be made available regardless of the user's location. However, the utility of the digital library is limited when users lack the means to access it.

Younger users: haves and have nots

The relevance of digital libraries for children is based largely on their ability to access that digital library and understand its content. 'Use of a digital library requires a computer attached to the Internet' (Arms, 2000). This revisits the issue of technology 'haves' and 'have nots'. Worldwide there are over 1.2 billion children between birth and age nine, 88% of whom live in less-developed countries. Adolescents comprise over a billion of the world's population, and 85% of those live in less-developed countries (UNICEF, 2000). 'More than [600 million children] live on less than $1 a day' (UNICEF, 2000). Young people in developing countries may not have regular access to computers and network connections, let alone digital libraries. For many children in developing countries, age-appropriate books written in their native languages are not readily available, and literacy cannot be assumed (Elkin, 1999; Patte and Hannesdóttir, 1984).

Australia, Denmark, Finland, Norway, Singapore, Sweden, Switzerland and the USA were reported as having more than 300 computers per thousand persons in 1999. Algeria, China, Egypt, Ghana, Guatemala, India, Indonesia, Jamaica, Morocco, Pakistan, Romania and Syria had fewer than ten personal computers per thousand persons (USA Department of Commerce, Census Bureau, 1999). Nua.com, which reports internet demographics and trends, estimated that there were 513.41 million internet users in August 2001. Of those users, 35% were in the USA and Canada, 30% in Europe, and 28% in Asia and the Pacific region (including Australia). Only 5% of estimated internet users were from Latin America, and less than 1% were from Africa or the Middle East (Nua, 2001a). Residents of developed countries who lack home access to computers or the internet can frequently gain access at their schools or public libraries (USA. National Telecommunications and Information Administration, 2000). However, when writing about South African public libraries, Hart (2000) stated that 'only seven of the 63 children's libraries had access to the Internet (all seven placed in historically white suburbs)'. A substantial change in infrastructure must occur before digital libraries become accessible to many younger users in less-developed nations.

In the UK 37% of households had internet access in 2001; 96% of primary and 99% of secondary schools were connected as well (Great Britain. Office for National Statistics, 2001). A 2001 report found that 5.6 million British children between ages seven and 16 have access to the internet (Nua, 2001b). In Austria, 80% of young adults between ten and 19 had internet access in 2001, using the internet to search for information and to send e-mail (Nua, 2001e). Almost 33% of Australian households had internet access in 2000 (Australian Bureau of Statistics, 2000). The vast majority of Australian schoolchildren had used computers, and 47% had accessed the internet, most doing so at school or at home, for educational purposes (Australian Bureau of Statistics, 2001). In Canada schools were the most commonly reported site for internet access (Clark, 2001).

However, even in more developed countries, children in marginalized groups may not have regular access to networked computers. All told, 69% of children in the USA have computers at home (Roberts et al., 1999). However, there is a relationship between computer access and marginalized status. Whereas 78% of Caucasian children report having access to a computer at home, only 55% of African-Americans and 48% of Hispanics can make that claim. Children whose parents were university educated and white were more likely to have used a computer on the previous day than children whose parents had not completed university or children whose parents were Black or Hispanic. Marginalized status also affects internet access. 'In August 2000, Whites (50.3%) continued to be the most likely to use the Internet, followed by Asian Americans and Pacific Islanders (49.4%), Blacks (29.3%) and Hispanics (23.7%).' Children from single-parent households are only half as likely to have internet access as children from two-parent households, and less than 20% of the poorest families were connected (USA. National Telecommunications and Information Administration, 2000). Australian rural and low-income households are less likely to have computers or internet access than metropolitan or higher-income households (Australian Bureau of Statistics, 2000). Poor children, rural children and children who do not come from the dominant class are simply less likely to access a digital library than their wealthier urban peers.

Age also plays a factor in children's use of the internet. In 2000 fewer than 10% of three-year-old American children used the internet. The demands of literacy and keyboarding skills put internet access beyond the intellectual capability of very young children. The older the child, however, the more likely that child was to use the internet, with the proportion of American internet users peaking at over 60% of 16-year-olds using the internet (USA. National Telecommunications and Information Administration, 2000). In Australia results were similar: 'the percentage of all [Australian] children accessing the Internet ranged from 8% for children aged 5 years to 76% for children aged 14 years' (Australian Bureau of Statistics, 2000). Of younger users aged seven to 16 in the UK, 83% were between the ages of 11 and 16 (Nua, 2001d), and in Canada parents of children ages 15 to 18 were twice as likely to report that their children had access to the internet (96%) as parents of children aged five to nine (48%) (Clark, 2001, p.7). The older a child grows, the more likely that child is to access the internet.

How do younger users use the internet?

Understanding how children and youth use the internet may help developers understand how they will use digital libraries. Those who do use the internet see it as a place to voice their opinions, a way to connect with other youth, a source of entertainment and a source of information. 'More than adults, children and youth see [the Web] as a place for self-expression' (Lazarus and Mora, 2000). A Pew report found a very strong communications role for the internet – 92% of American teenagers have sent e-mail, 74% have sent an instant message, and 55% have visited chat rooms (Lenhart, Rainie and Lewis, 2001). Australian children aged 12 to 14 also use the internet's communications power: 64% have used the internet for e-mail and chat (Australian Bureau of Statistics, 2001).

Entertainment use of the internet plays a substantial role in its attraction to younger users. In the UK 'half [of all users under 17] visited music and literature sites and 40 per cent logged onto games sites. Downloading music and video clips, and using instant messaging are also popular among this age group' (Nua, 2001d). A study of minority and low-income internet users in the USA found that the informational purpose of the internet took a back seat to entertainment. 'Young people in our study see the Internet primarily as a place for gaming and participating in interactive communities with kids all over the world' (Lazarus and Mora, 2000). These young people also wanted convenience in their internet experience. 'Many young people we talked to want more centralized spaces where they can participate in a variety of ways from one portal. It would work best for them if one site contained games; downloadable plug-ins; tips and strategies; e-mail; user profiles; and links to other game environments' (Lazarus and Mora, 2000). Younger users in Los Angeles, aged six to 14, felt similar: they wanted animation, interactivity and rapid down-loading (Enochsson, 2001). The users surveyed by Lazarus and Mora (2000) wanted 'fast moving imagery and sound'. Slow download times hinder young people's use of the internet.

Educational use of the internet has not been completely elimi-nated, however. The primary reason why Australian young people access the internet is for educational use (Australian Bureau of Statistics, 2001). The vast majority of US teenagers, 94%, state they use the internet for school research (Lenhart, Rainie and Lewis, 2001). Moreover, a majority of American young adults think that 'increased access to computers and the Internet' would improve their schools – 26% said it would improve schools 'a little', but 56% said it would improve schools 'a lot' (Horatio Alger Association, 2001). The internet has not yet lost its mystique as an information resource.

Linguistic and developmental barriers to younger users

Beside the barrier of having limited access to a traditional library or the internet, children and young adults may also face a barrier in communicating their information needs in speech or via a keyword search. They may not be able to articulate their needs with the preci-sion that adults can, as a result of limited vocabularies and life expe-riences. The Science Library Catalog project, developed at the University of California – Los Angeles in 1989,

> was designed to minimize the known difficulties children have with existing online catalogs (e.g., spelling, typing/keyboarding, alphabetiz-ing, Boolean logic) and to build on their skills and abilities (e.g., brows-ing, recognizing relevant topics, navigating hierarchical displays, using a mouse). (Hirsh, 1997)

However, even when using this graphical search interface, children who knew more about their subject performed better at finding resources than children who knew less (Hirsh, 1997). And although young adults in secondary school would be assumed to have increased domain knowledge compared with children, Chen (1993) reports that students searching the online catalogue also have difficulty thinking of keywords.

Studies of children's information needs show that, in many cases, those needs are imposed upon children by adults, rather than being needs felt by children themselves (Gross, 2000). Children may seek information to satisfy a school requirement or answer a question from a parent. In doing this, they have no context in which to understand the information problem. A study of New Zealand Year 6 students searching a library catalogue found that the children did not always understand the relationship between catalogue records and books. 'The efficacy of the search depends on the individual's ability to match keywords to information sources, to see relationships between differing aspects of the topic' (Moore and St George, 1991). Based on her study of Swedish children aged nine to 11, Enochsson (2001) speculates that there is a 'development of children's awareness about Web pages'. As children think more reflectively about the content presented on a web page, they also become more critical of the veracity of that content. Enochsson suggests that these children are moving away from concrete thinking and moving toward abstract thinking, and that practice with web resources will help them make this leap more easily. Nonetheless, children are as subject to information overload as adults (Akin, 1998).

Digital libraries for younger users

Historically, digital libraries have not been developed for youth. They were originally produced to meet the needs of corporations, and for the scientific community to share information. Consequently, digital library research is based largely on populations of educated adults who share a similar culture or perspective (Bruce and Leander, 1997). Compared with a group of similarly enculturated professionals or similarly educated academics, the world population of two billion young people is dramatically heterogeneous. It is composed of children and young adults from a variety of backgrounds, with a variety of linguistic skills and learning needs. Nonetheless, in their state of the world's children report, UNICEF (2000) insists that education programmes 'must . . . use information and communication technologies to reduce disparities in access and quality'.

Bruce and Leander (1997) maintain that, for many in the school community, the first exposure to the learning potential of digital libraries has come via the CD-ROM encyclopedia. The CD-ROM encyclopedia carries information on a variety of subjects, meeting the diverse information needs of students and teachers. It makes use of multimedia features such as sound and video clips, enhancing information delivery potential. The CD-ROM encyclopedia provides information directly to the reader. Its content is structured and organized consistently, and this content is owned and controlled by the publisher.

The development of digital libraries in some ways echoes the development of youth services librarianship. Analysing the term 'digital library', Schwartz found that, even among their creators, the definition of what constitutes a digital library was not consistent. However, much like specialized youth services, a common trait across digital libraries is that they are designed to 'serve a defined community' (Schwartz, 2000). Youth services librarianship is also designed to serve a defined community. The foundation upon which youth services librarianship has built itself is, of course, children's materials, and more specifically, children's books. The acquisition of children's materials provides an impetus for the hiring of children's librarians and the establishment of children's rooms (Jenkins, 2000). 'Much of the impetus for digital library research and development was content' (Marchionini, 1998). The Digital Libraries Initiative web page says, 'Digital Libraries basically store materials in electronic format and manipulate large collections of those materials effectively'. It also indicates that the Digital Libraries Initiative seeks to 'dramatically advance the means to collect, store, and organize information in digital forms, and make it available for searching, retrieval, and processing via communication networks' (Digital Libraries Initiative, 1998). These definitions of a digital library focus almost exclusively on the information-bearing object.

Because adult-serving digital libraries are designed for the rapid, efficient provision of information, specialized programmes are not common. In this area, youth-serving digital libraries have taken the forefront with interactive games, communications with other youth, and special author presentations (Bussman and Mundlechner, 1998). The provision of specialized personnel via digital interface is also becoming more common. The Virtual Reference Desk, a project supported by the US Department of Education, connects users to a variety of 'Internet-based question-and-answer services that connect users with experts and subject expertise' (*About VRD*, 2000). The Virtual Reference Desk and its compilation of 'AskA' services were designed to benefit the educational community, although the general public are invited to participate. Another programme, the Collaborative Digital Reference Service, has developed international collaboration to '[provide] professional reference service to researchers any time, anywhere' (Library of Congress, accessed 2002). Services like this connect the digital library to its user in a way that merely providing access to information cannot.

Current research

Children's interactions with electronic media have been studied extensively, building a base of knowledge that can be drawn upon for digital library creation. In 1995 Jacobson wrote, 'electronic information retrieval products have not been designed with the developmental needs of children in mind'. In some cases this is still true, though today's digital libraries for youth make a concerted effort to meet those needs. The University of Maryland currently has a research grant from the Digital Libraries Initiative-2 to study how digital libraries can provide for the intellectual and developmental needs of youth. 'It is common for developers of new technologies to ask parents and teachers what they think their children or students may need, rather than ask children directly' (Druin, n.d.). However, faculty at the University of Maryland work directly with children to develop a product that meets their needs. By working directly with children the digital library developers at the University of Maryland have developed a deep understanding of the way children interact with computerized information resources.

This deeper understanding has enabled them to develop a visual query interface, allowing children to search using graphical representations of known information. Zoomable User Interfaces (ZUIs) are another product of this project. These ZUIs allow children to zoom in on individual elements of a story, rather than being restricted to the linear story format of most digital library stories. And because children enjoy working collaboratively, the project is developing digital library interfaces that support multiple users. Children and teenagers in US libraries can be seen clustering around internet terminals, and to them using these resources is a joint effort (Lenhart, Rainie and Lewis, 2001). Technology developed by this project may give these children each an active role in navigating digital libraries.

A brief look at some digital libraries for children
The Internet Public Library Youth Division
www.ipl.org/youth/

The Internet Public Library is a service of the University of Michigan's School of Information. There are IPL mirror sites in Argentina and Japan; however, the interface for both these sites is English, which limits the site's utility for non-English speakers. IPL bills itself as 'the first public library of and for the Internet community'. The IPL-Youth interface is structured like a web directory, providing the user with a list of websites. This eliminates the need for the ability to refine searches carefully and retrieve only the most relevant resources, although children looking at the annotations for these websites may recognize concepts they are studying and investigate further.

The websites available via IPL-Youth have been chosen carefully on the basis of relevance and age-appropriateness, and, like a traditional library, IPL-Youth makes its selection policy available for its clients. The Internet Public Library does not own this content; it merely provides links to it. Some would argue that this eliminates it from consideration as a digital library (Peek, 1998). Another service provided by IPL-Youth is the ability to submit questions to volunteer IPL librarians and receive an answer. However, the Ask A Question form (www.ipl.org/youth/refform.html) is very text-oriented, and may be daunting for younger children or children with reading disabilities.

The IPL Youth Division serves primarily as an information resource. Although it provides links to games, IPL-Youth does not have interactive or animated content. Graphic images are used to identify categories, but these are basic icons that often do not completely represent the category they are illustrating. For instance, the Math Whiz category is represented by a wizard's hat. The child seeking mathematics information may not connect a wizard's hat with the concept of mathematics. The child or young adult using this site must be able to read, and to read English. Younger users who think verbally, in words, are more likely to benefit from this resource than children who think visually, in pictures.

America's Library

www.americaslibrary.gov/

America's Library is a project of the US Library of Congress, with the intention of '[putting] the story back in history'. The project was launched in 2000 and in its first month of operation had received over six million hits (New Web, 2000). The photographs and images displayed here are owned by the Library of Congress and are a subset of what is available in the American Memory Digital Library (http://memory.loc.gov/). America's Story does not try to be a comprehensive source of historical information. The collection is limited to American history, and only brief information on specific subjects is provided.

The initial page for America's Story features bright colours, animated graphics and relatively little text. Each page is consistently presented with the America's Story logo at the top, helping the user to maximize the sense of place or uniformity provided by this digital library. Five modules are available in America's Story, plus two feature stories which change periodically. The brief text vignettes are supplemented by video, audio and historical pictures. Each content page has a balance of graphics and text, preventing the user from becoming overwhelmed by one or the other. The interactive content on the America's Story Digital Library includes scavenger hunts, games, and the Jammin' Jukebox, which allows users to select among ten audio and video pieces to play.

America's Story does not offer reference services for younger users, nor does its parent institution, the American Memory Digital Library. However, the Library of Congress also makes available an American Memory Learning Page (http://lcweb2.loc.gov/ammem/ndlpedu/), designed to help K-12 educators use the American Memory Digital Library in their classrooms. Entertainment has been a significant goal for America's Story's developers; the advertising campaign's slogan is 'there is a better way to have fun with history . . . . Log On. Play Around. Learn Something'(New Web, 2000). While the child who does not read may find some enjoyment in the interactive media and pictures, English language literacy is essential for those children hoping to maximize their informational experience at America's Story.

StoryPlace

www.storyplace.org

StoryPlace is provided by the Public Library of Charlotte and Mecklenberg County (PLCMC) in North Carolina. StoryPlace is the product of the library's children's services staff and web developers. The site opened in 2000 with the Preschool Library. The library joined with Smart Start of North Carolina, an early childhood education advocacy group, to provide this content. *Computers in Libraries* reported in 2001 that this site was 'co-branded' with internet search engine Lycos (Public Library, 2001), but at the time of writing co-branding was not apparent.

StoryPlace content is made available in Spanish and English, and the interface makes considerable use of graphics. It features two modules, the Preschool Library/La Biblioteca Pre-escolar, and the Elementary Library/La Biblioteca Infantil. The Preschool Library module has several themes, and each theme has an online story, an online game activity, a 'take-home' activity for children to make at home, an annotated reading list, and a parent activity sheet featuring songs and fingerplays. Themes in La Biblioteca Pre-escolar did not have the parent activity sheet, and the reading lists were not as extensive. In both languages the online stories are read aloud; therefore, even pre-literate children can use the Preschool Library. Basic mouse skills are reinforced through the online activities, and keyboarding skills are not necessary.

The Elementary Library has fewer modules than the Preschool Library, and no online games. The narrated stories have more text in them, making basic literacy a necessity, but the user can personalize the stories by entering names for the characters. StoryPlace provides a link to the English-only Book Hive (www.bookhive.org/), another PLCMC service, where parents and children can read reviews of books and add their own comments. Other than the reviews in the Book Hive, librarians do not interact with their StoryPlace readers.

While consistent in design, the pages are graphics heavy and take some time to load on slower connections. A list of system recommendations is available for maximum enjoyment of the site, including a connection speed of 56K or better, Flash 5.0, and a Pentium 200 processor. These requirements place StoryPlace out of the hands of the majority of US users – in 2000, only 9% of USA households had broadband access to the internet (Nua, 2001c)

CHILIAS Biblioteca Virtual
http://chilias.diba.es

CHILIAS (CHIldren's Library – Information – Animation – Skills) was designed to serve European children aged nine to 12. This virtual children's library was funded by a grant from the European Commission's Telematics Applications Programme in 1996, and was completed in 1998 (Bussman and Mundlechner, 1998). Test sites were established in six countries: Finland, Germany, Greece, Portugal, Spain and the UK. At the time of writing sites were accessible in Finland, Germany, Greece and Spain. The Greek site (www.haef.gr/chilias/) has an English translation; however, no British CHILIAS site is available. Much of the content available on the Greek version of CHILIAS was created by Greek children themselves, but the Greek site has been stagnant since 2000.

The Spanish site is in Catalan, a language spoken in some regions of Spain, including Barcelona, where the site is based. After the initial CHILIAS project was completed in 1998, the Spanish CHILIAS site was redesigned to integrate library and web services. In order to accomplish this, the site linked to the shared catalogue of the three participating libraries. This digital library provides information via a list of approved websites, which includes advice on citing sources and judging the accuracy of information provided on the web. Most of the websites to which the site links are in Spanish or Catalan. In-depth information on specific subjects is provided in the Temas (Themes) module. For instance, under the broad subject heading of 'animals', it provides websites, a searchable database, and featured animal of the month. The Biblio i ciutat (library and city) module allows users to learn more about the three Spanish libraries involved in this effort and their communities.

Entertainment and self-expression feature in the Tallers (Studios) module. The Taller de contes (story studio) challenges children to make up stories incorporating several unrelated pictures. The games page features interactive memory and mathematics games, while the painting page features pages to colour. The Grups d'opinia (chat forums) give users a place to debate issues such as how the monetary change to the Euro would affect them. There is an option to send a message to the creators of the digital library and photographs of the children's services staff; otherwise, this service does not appear to provide personalized reference services. At present it seems to be inactive.

Conclusion

A digital library for adults is a computer-accessible content store-house. In traditional libraries children's services promote a love of reading and learning. Younger users want entertainment, interactivity and multimedia. Digital libraries for younger users do not presently incorporate all of these features. StoryPlace does have interactive content and high entertainment value, and serves to provide a love of reading and learning, but its content supply is fairly limited. America's Library also has interactive, entertaining content, but is not designed for children with limited literacy skills. While the love of learning might be promoted, the love of reading needs to be developed beforehand, and the content, while extensive, is limited. The Youth Division of the Internet Public Library has a considerable supply of content, but is not entertaining, interactive or particularly well indexed. The CHILIAS Biblioteca Virtual is perhaps the most developed of the digital libraries for youth. However, content is limited and development seems to have ceased. Some components of the library are interactive, while others are reflective.

In 1989 Adele Fasick wrote that 'one of the most obvious changes in children's libraries has come about because of changing media'. Computerized catalogue systems, CD-ROMs and interactive children's software made huge impacts when they were introduced. In 2002 a great change in children's services will come about as a result of digital libraries for youth. By providing digital library services that recognize that children think and learn differently from adults, today's digital library developers are following in the footsteps of the pioneers of children's librarianship. Originally designed for adults, digital libraries were dedicated to rapid information transfer. A new generation of children who have grown up with the internet and online gaming, who expect content to be entertaining as well as informative, and who have different linguistic and developmental abilities, will expect different types of digital libraries.

References

About VRD (2000)
   http://vrd.org/about.shtml
   (accessed 23 February 2002)
Akin, L. (1998) Information overload and children: a survey of Texas elementary school students, *School Library Media Quarterly Online*, **1**, available at
   www.ala.org/aasl/SLMQ/overload.html
   (accessed 14 February 2002)
American Association of School Librarians and Association for Educational Communications and Technology (1988) *Information power: guidelines for school library media programs*, Chicago: American Library Association; Washington, DC: Association for Educational Communications and Technology.
Arms, W. Y. (2000) *Digital libraries*, Cambridge, MA: MIT Press.
Australian Bureau of Statistics (2000) *AusStats: 8147.0 ABS. Increasing use of the internet and home computers*, available at

www.abs.gov.au/ausstats/abs@.nsf/0/
8A1031CEF42CB4E6CA25694500804435?Open&Highlight=
0,computers
(accessed 15 February 2002)

Australian Bureau of Statistics (2001) *AusStats. 4901.0 ABS. Children's participation in cultural and leisure activities,* Australia, available at www.abs.gov.au/ausstats/abs@.nsf/0/
0B14D86E14A1215ECA2569D70080031C?Open&Highlight=
0,computers,libraries,internet
(accessed 15 February 2002)

Bredsdorff, A. (1978) Denmark. In Ray, C. (ed.) *Library service to children: an international survey,* New York: K.G. Saur, 29–37.

Bruce, B. C. and Leander, K. M. (1997) Searching for digital libraries in education: why computers cannot tell the story, *Library Trends,* **45** (4), 746–70.

Bussman, I. and Mundlechner, B. (1998) CHILIAS – The European Virtual Children's Library on the internet: a new service to foster children's computer literacy. Paper presented at the 64th IFLA General Conference, Amsterdam, available at www.ifla.org/IV/ifla64/043-113e.htm
(accessed 15 February 2002)

Chen, S. (1993) A study of high school students' online catalog searching behavior, *School Library Media Quarterly,* **23** (1), 33–9.

Clark, W. E. (2001) Kids and teens on the net, *Canadian Social Trends,* 6-10, available at www.statcan.ca/english/kits/pdf/social/net2.pdf
(accessed 17 February 2002)

Digital Libraries Initiative (1998) *National synchronization* http://dli.grainger.uiuc.edu/national.htm
(accessed 3 February 2002)

Druin, A. (n.d.) *University of Maryland's Digital Libraries for Children,* available at www.cs.umd.edu/hcil/kiddiglib/project.html
(accessed 25 February 2002)

Elkin, J. (1999) Informal programmes to support reading and libraries in developing countries, *The New Review of Children's Literature and Librarianship*, **5**, 55–84.

Enochsson, A. (2001) Children choosing web pages, *The New Review of Information Behaviour Research: Studies of Information Seeking in Context*, **2**, 151–66.

Great Britain. Office of National Statistics (2001) *UK 2002: The official yearbook of Great Britain and Northern Ireland*, London: The Stationery Office, available at
www.statistics.gov.uk/downloads/theme_compendia/UK2002/UK2002.pdf
(accessed 16 February 2002)

Gross, M. (2000) The imposed query and information services for children, *Journal of Youth Services in Libraries*, **13** (2), 10–17.

Hannesdóttir, S. K. (1978) Iceland. In Ray. C. (ed.) *Library service to children: an international survey*, New York: K. G. Saur, 66–71.

Hart, G. (2000) A study of the capacity of Cape Town's children's librarians for information literacy education, *Mousaion*, **18** (2), 67–84.

Hirsh, S. G. (1997) How do children find information on different types of tasks? Children's use of the Science Library Catalog, *Library Trends*, **45** (4), 725–45.

Horatio Alger Association of Distinguished Americans (2001) *The state of our nation's youth: 2001-2001* (2001) Alexandria, VA: Horatio Alger Association of Distinguished Americans, Inc. available at
www.horatioalger.com/pdfs/state01.pdf
(accessed 20 October 2001)

Hume, H. (1978) Australia. In Ray, C. (ed.) *Library service to children: an international survey*, New York: K. G. Saur, 9–16.

Jacobson, F. F. (1995) From Dewey to Mosaic: considerations in interface design for children, *Internet Research: Electronic Networking Applications and Policy*, **5** (2), 67–73.

Jenkins, C. (2000) The history of youth services librarianship: a review of the research literature, *Libraries and Culture* **35** (1), 3–140.

Klass, E. and Perumbulavil, V. (1978) Singapore. In Ray, C. (ed.) *Library service to children: an international survey*, New York: K. G. Saur, 114–23.

Lazarus, W. and Mora, F. (2000) *Online content for low-income and underserved Americans: the digital divide's new frontier*, Santa Monica, CA: The Children's Partnership.

Lenhart, A., Rainie, L. and Lewis, O. (2001) *teenage life online: the rise of the instant-message generation and the internet's impact on friendships and family relationships*, Washington, DC: Pew Internet and American Life Project, available at www.pewinternet.org/reports/pdfs/PIP_Teens_Report.pdf (accessed 17 February 2002)

Library of Congress. Collaborative Digital Reference Service www.loc.gov/rr/digiref/about.html (accessed 23 February 2002)

Makafu, O. L. (1978) Tanzania. In Ray, C. (ed.) *Library service to children: an international survey,*. New York: K. G. Saur, 124–31.

Marchionini, G. (1998) Research and development in digital libraries. In Kent, A. and Hall, C. M. (eds) *Encyclopedia of library and information science,* vol. 63, supplement 26, New York: Marcel Dekker, 259–79.

Moore, P. A. and St. George, A. (1991) Children as information seekers: the cognitive demands of books and library systems, *School Library Media Quarterly*, **19**, 161–8.

New web site for kids and families registers more than 6 million hits since debut (2000) *News from the Library of Congress*, 30 May, available at www.loc.gov/today/pr/2000/00-085.html (accessed 24 February 2002)

Nua Internet Surveys (2001a) *How many online?*, available at www.nua.com/surveys/how_many_online/index.html (accessed 15 February 2002)

Nua Internet Surveys (2001b) *Most kids now online in UK,* available at www.nua.com/surveys/index.cgi?f=VS&art_id=905357089&rel=true (accessed 15 February 2002)

Nua Internet Surveys (2001c) *Two in five households to be hi-speed by 2005,* available at www.nua.com/surveys/index.cgi?f=VS&art_id=905357311&rel=true (accessed 25 February 2002)

Nua Internet Surveys (2001d) *Under 17 net audience in UK exploding,* available at www.nua.com/surveys/index.cgi?f=VS&art_id=905357654&rel=true (accessed 14 February 2002)

Nua Internet Surveys (2001e) *Young Austrians embrace the net,* available at www.nua.com/surveys/index.cgi?f=VS&art_id=905356634&rel=true (accessed 15 February 2002)

Omolayole, O. O. (1978) Nigeria. In Ray, C. (ed.) *Library service to children: an international survey,* New York: K. G. Saur, 102–7.

Onaga, I. G. B. (1978) Brazil. In Ray, C. (ed.) *Library service to children: an international survey,* New York: K. G. Saur, 17–20.

Patte, G. (1978) France. In Ray, C. (ed.) *Library service to children: an international survey,* New York: K. G. Saur, 43–50.

Patte, G. and Hannesdóttir, S. K. (eds) (1984) *Library work for children and young adults in the developing countries/Les enfants, les jeunes et les bibliotheques dans le pays en developpement: Proceedings of the IFLA/UNESCO pre-session seminar in Leipzig, GDR, 10–15 August 1981,* Munich: K. G. Saur.

Peek, R. (1998) Miss Web Manners on digital libraries, *Information Today,* **15** (7), 36.

Public library joins with Lycos for StoryPlace (2001) *Computers in Libraries,* **21** (4), 12.

Ray, C. (1978) United Kingdom. In Ray, C. (ed.) *Library service to children: an international survey,* New York: K. G. Saur, 132–9.

Ray, S. G. (1979) *Children's librarianship,* London: Clive Bingley.

Roberts, D. F. et al. (1999) *Kids & media @ the new millennium: a*

comprehensive national analysis of children's media use, a Kaiser
Family Foundation Report, available at
www.kff.org/content/1999/1535/KidsReport%20FINAL.pdf
(accessed 24 October 2001)

Schwartz, C. (2000) Digital libraries: an overview, *Journal of Academic Librarianship*, **26** (6), 385–93.

UNICEF (2000) *The state of the world's children 2000*, New York:
UNICEF, available at
www.unicef.org/sowc00/pdf.htm
(accessed 20 October 2001)

USA. Department of Commerce. Census Bureau (1999) *Statistical abstract of the United States: 1999*, Washington, DC: Government
Printing Office, available at
www.census.gov/prod/99pubs/99statab/sec01.pdf
(accessed 22 October 2001)

USA. National Telecommunications and Information Administration
(2000) *Falling through the net: toward digital inclusion*, available at
www.ntia.doc.gov/ntiahome/fttn00/contents00.html
(accessed 2 February 2002)

Vargha, B. (1978) Hungary. In Ray, C. (ed.) *Library service to children: an international survey*, New York: K. G. Saur, 59–65.

Willett, H. G. (1995) *Public library youth services: a public policy approach*, Norwood, NJ: Ablex Publishing Corporation.

Part 4

REFERENCE SERVICES

8

Web-based reference services: design and implementation decisions

Stephen M. Mutula

Introduction

Today many libraries are building home pages on the world wide web, while others are in transition from the traditional walled to the digital library environment in response to the demands of increasingly sophisticated users (Corbin and Coletti, 1995). The ubiquity of the internet combined with the benefits that accrue from using web-based information resources has seen a growing number of libraries reorganize their services to take advantage of these innovations. Libraries are motivated by the benefits they can enjoy in cyberspace: web-based resources can be stored inexpensively in compact form; digital documents can be searched in seconds, and their content reshaped to the reader's needs; the multimedia aspects of web resources facilitate use of the learners' full sensory capabilities; and web-based services can be customized to individual user's needs. Additionally, web reference services provide users with the convenience of accessing information in their own time, saving them travelling costs and time. The provision of these services is not constrained by the traditional opening hours but can be offered on a 24-hour, seven-days-a-week basis (known as 24/7).

Despite the attractions associated with web-based services, the library must have the appropriate infrastructure in place – computers, software applications and telecommunications. Lack of standards (both in the form of common technical standards and collection policies) for organizing information on the web poses great challenges to librarians. The standards for organizing web-based resources are still in the early stages of development, and librarians are forced to utilize standards for print resources that were not designed for electronic resources. Additionally web-based information resources are volatile in the sense that they may be moved from one site to another or may be removed altogether from the web. Further, provision of remote reference services without physical contact with the user imposes constraints on effective interactions. The richness of face-to-face communication in relation to body language cannot be exploited to help understand the user and his or her requirements.

It is still early days in the development of web-based reference services, and there is not yet any national or global co-ordination in this field. This imposes constraints on quality, and often information obtained from the web lacks the authority and authenticity associated with print resources. Further, the development of web-based information resources is a complex process involving multiple dimensions that need to be balanced, such as user–systems interactions, standardization of provider infrastructures, issues of copyright and intellectual property, etc. The provision of web-based information services also demands that information providers acquire additional skills to act as intermediaries.

As more libraries move towards providing reference services in a digital environment, the improved access to remote library collections is making the use of electronic information resources more feasible and more attractive. Online catalogues are evolving into gateways to online resources, not only to books and serial titles held locally but to other collections elsewhere and, through indexing and abstracting services, to articles in periodicals. Libraries have much at stake if they cannot evolve quickly enough to take advantage of the enormous volume of information on the web.

Evaluation and selection of web resources

Evaluation of web resources precedes their selection for inclusion in local collections' OPACs or to be pointed at by an OPAC. Evaluation helps ensure that the resources selected are relevant and of good quality.

As with conventional information resources, it is important when evaluating the resources to know who the author is and whether this person is authoritative in the field. If the resource is being sponsored by an organization, it is important to establish the legitimacy of such sponsorship – does, for example, the sponsor have a vested interest in the content that might prejudice its objectivity? The Universal Resource Locators (URLs) may also provide some hint of the credibility of the source. A source with '~' in the URL, for instance, usually indicates that the site belongs to an individual rather than an organization and therefore its credibility may be more questionable (although not necessarily so). It is also essential when evaluating web resources to consider whether the material is protected by copyright, and, if so, the name of the copyright holder and contact details for the purpose of seeking permission to use the resource.

It is important that the information resource selected is free from grammatical, spelling and typographical errors. For those sources that contain charts, graphs or statistical data, these need to be clearly drawn and labelled for ease of reading. The importance of currency cannot be over-emphasized, as this is supposed to be one of the virtues of the web. The currency of the source can be determined in part by establishing when the page was written, when it was first placed on the web or when it was last revised.

Consideration should also be given to whether the resource is appropriate for the level of the expected users. Highly technical sources, for example, should be selected for those users who are experts in the field, while for non-technical users sources should be more elementary. It is crucial to evaluate the ease with which information can be retrieved from the website, because the more difficult the retrieval, the less satisfied the user in most instances. Further, the sources selected should be comprehensive in their coverage to save the user time by making it less necessary to move from one resource to another before acquiring the needed information (Cornell University Library, 2001).

In a similar vein, a web-based reference service must be evaluated regularly to ensure that it meets user needs. One common way of evaluating the service is to ask the users how they feel about the service and what they think needs to be done to improve service delivery. Similarly, the extent of web use by the target group can provide an indication of the appeal of the web page content (Garlock and Piontek, 1996). If the target group is not using the web page, one conclusion may be that the content is not meeting the information needs of that group. Usage can be measured in a number of ways, one of the most attractive at present being the use of web page trackers that are provided free on the web. One example is the tracker from (www.eXTReMe www.extreme-dm.com/tracking/?npt), which provides detailed reports on the number of unique visits per day, week and month; time of day the web page was accessed; user location by country, etc. Other ways to evaluate the service include putting in place performance indicators against which service can be measured, such as the availability level of the service over a period of time, what external evaluators say about the service, etc.

Organization and management of web resources

For electronic resources to be easily accessed and used they need to be organized in a logical, easily understandable manner. In general the classification should also be simple and intuitive for it to appeal to the users. The classification of web resources involves mapping content that is already on a website and devising a structure for the site to facilitate access to the resources. Most organizations tend to have a pre-existing classification system that may be understood by its employees but not its clients (see *University of Wolverhampton. School of Computing and Information Technology* www.scit.wlv.ac.uk/wwwLib/). It is important that a consistent classification be employed across the website to make it easier for managing and creating a familiar environment for visitors. Today traditional cataloguing methods are being applied to the description of web resources, and MARC is also being used increasingly for the cataloguing of web resources. Many integrated library systems are able to apply MARC standards to support access to networked resources and the discovery and retrieval of information.

There are other emerging standards for organizing material on the web, such as Text Encoding Initiative (TEI), which is being used to create descriptive metadata. And there are other initiatives, such as the OCLC InterCat Project, aimed at creating implementing, testing and evaluating a searchable database of USMARC format bibliographic records for internet-accessible materials (IFLA, 1999). Other initiatives include the Draft Interim Guidelines for Cataloguing Electronic Resources by the Library of Congress, based on a common conceptual context and a common terminology. The guidelines include:

- policies on when to use multiple records and when to use a single record
- conventions developed especially for the single-record technique
- directions for indicating the existence of other physical formats
- directions related to electronic location; directions relating to co-locating records and linking them
- directions relating to identifying records related to specific projects (Library of Congress, 1997).

Other developments in organizing web resources include consideration of the question of whether to use a single record or separate records for online and print versions. Currently, no single standard has been adopted, but there are many advocates for the one-record approach; as a result, the current working trend is to adopt the single-record approach for multiple formats. The decision to catalogue an electronic resource as a collection, a collection with analytics, or as individual titles is left to individual libraries to decide as a local choice. Ease of access to the individual titles, full online content and stability of the titles represented will influence this decision. Depending on whether the library decides to catalogue at the collection level or the title level, the URL from the bibliographic record will point to a database of collected titles, an individual journal, a single database or a monographic title.

Because of the problems of instability and short life of web resources, there are several efforts aimed at addressing this problem. For example, OCLC is assigning Persistent Uniform Resource Locators (PURLs) to ensure that, even if the site is moved, the link will still point to the new location. Similarly, the University of Chicago uses a 'handle' system, which provides a means of naming resources in a way that persists over changes of location and ownership, while other libraries use bookmark files as a quick way to keep track of websites.

The classification of web resources is being achieved through automatic methods, which involve 'teaching' systems to recognize documents belonging to particular classification groups. Figure 8.1 shows how automatic classification is achieved in the WWLib.

In this approach the Spider retrieves documents from the web, while the Indexer receives web pages from the Spider, stores a local copy, assigns to it a unique accession number and generates a new metadata template. The Indexer distributes local copies to the Analyser, which analyses pages provided by the Indexer for embedded hyperlinks to other documents. The Classifier and Builder add subsequent metadata generated by the Spider to the assigned metadata template. The Classifier analyses pages provided by the Indexer and generates DDC class marks, while the Builder analyses pages provided by the Indexer and outputs metadata which is stored by the

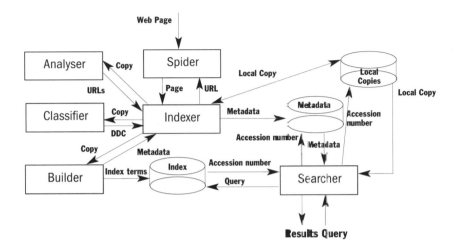

Fig. 8.1   *Overview of the new WWLib architecture*
Source: University of Wolverhampton. School of Computing and Information Technology

Indexer in the document's metadata template and is also for the index database that will be used to associate keywords with document accession numbers. The Searcher accepts query strings from the user, uses them to interrogate the index database created by the Builder, uses the resulting accession numbers to retrieve the appropriate metadata templates and local document copies and then combines all this information to generate detailed results, ranked according to relevance to the original query.

Web-based resources and services

There are different kinds of web-based reference resources and services including gateways, search engines, portals, electronic journals, subject directories and online databases. These resources overlap considerably in the type of information they cover, and sometimes it is difficult to distinguish between some of them.

Gateways

A gateway is defined as a facility that allows easier access to network-based resources in a given subject area (Dan, Welsh and Martin, 1998). Gateways index lists of links and provide a simple search facility. They also provide a much-enhanced service via a system consisting of a resource database and various indexes, which can be searched through a web-based interface. Within a gateway, each entry in the database contains information about a network-based resource, such as a web page, website, mailing list or a document (see ADAM Consortium). Information provided by gateways is catalogued by hand, allowing keywords to be added to the record, which enables more relevant results to be retrieved and offers the opportunity to develop thesaurus-based searching. Gateways cover a wide range of subjects, though some areas, such as music and religious studies, currently lack subject gateways. Examples of some well-known gateways include the Internet Public Library (IPL), the Bulletin Board for Libraries (BUBL), National Information Services and Systems (NISS) (www.niss.ac.uk/subject/index.html) and Social Science Gateway (SOSIG) (www.sosig.ac.uk/).

Portals

Portals are either commercial or free web facilities that offer information services to a specific audience. The information provided by portals ranges from web search to communication in the form of chat and e-mail to news and reference tools as well as shopping information. There are different kinds of portals: consumer (or horizontal), vertical and enterprise. Consumer portals are aimed at consumer audiences and typically offer free e-mail, personal home pages, sports scores, stock reports, instant messengers, auctions, chat and games. Examples of consumer portals include Yahoo!, MSN and AOL. Vertical portals, exemplified by VerticalNet, target a specific audience, such as a particular industry, and offer many of the consumer portal features. Frequently the user must be a member of a particular group or organization to access the vertical portal.

Enterprise portals on the other hand are similar to consumer portals, but they are offered only to corporations or similar organizations. Examples of Enterprise portals include Epicentric and Corporate Yahoo! Examples of portals that offer focused services include Microsoft SharePoint, MySAP and Oracle Portal. Microsoft SharePoint gives users the ability to organize information, readily access that information, manage documents, and collaborate efficiently, all in a familiar, browser-based and Microsoft Office-integrated environment. MySAP delivers a comprehensive e-business platform designed to help companies collaborate regardless of their industry or network environment. Oracle Portal helps to develop and deploy enterprise web portals, employing interfaces for end-users and administrators to organize and publish information to better serve customers, drive revenue, and improve employee efficiency (Traffick.com, 2001).

Search engines

Search engines are spider-crawling software that indexes websites into a massive database that a user can search using keywords. Examples of well-known search engines include Google and AltaVista. A variant form of search engine is the meta-search engine, which goes behind the scenes to gather the top search results based on keyword searches from various search engines, and return the results with the source identified. Examples of meta-search engines include Mega Spider, Metacrawler, Ixquick and Mamma. The other type of search engine is the paid inclusion engine, which in addition to featuring normal search results allows sites to bid on their placement by paying a certain amount every time a searcher clicks on their link. Examples of such search engines include Yahoo!, GoTo.com, Sprinks and Kanoodle.

Search engines index anything and everything submitted to them, and this often creates unnecessary bulk in web collections. There is also lack of human indexing, and the search engine technology requires people with the requisite skills to use them effectively. Usually search engines do not focus on the subject or audience but instead collect information on all subjects for all people and are therefore often a source of frustration. Collecting for a particular audience's needs, and reducing the subject matter of the collection, can achieve reductions in the number of irrelevant links.

Subject directories

Subject directories differ from search engines in that search engines are populated by robots that find and index sites, whereas humans making editorial decisions populate subject directories. Subject directories are classified collections of websites and are arranged in hierarchical order, topically or alphabetically. Subject directories basically index home pages of sites and can be classified as general, academic, commercial or portal. A directory when searched lists the location of sites that fit the description specified but contains no detailed information about the contents of the sites. Among the well-known subject directories are the Argus Clearinghouse (www.clearinghouse.net) and Yahoo! (www.yahoo.com). The strengths of subject directories include the relevance of materials located, effectiveness in finding general information and the relative high quality of content. The weaknesses of subject directories lie in the fact that they lack depth in their coverage of the subjects, often contain many dead links because they are not automatically updated and do not compile databases of their own but merely point to existing ones.

Electronic journals

Today many journals are available electronically – some are full text and often are not very different in quality from the printed journals, while others contain only bibliographic information with abstracts. The handling of electronic journals requires the same intellectual and editorial input as do the printed journals, and only their production and distribution are different. One major advantage of electronic journals is that they are constantly updated and easy to access. A disadvantage is that copyright law is easy to breach. Electronic journals come in a variety of formats and delivery methods such as bitmaps, PostScript, PDF (Portable Document Format) as well as ASCII (American Standard Code for Information Interchange), SGML and HTML. They might be delivered on CD-ROM, through e-mail, or through networks (Wusteman, 1996).

The acceptance of electronic journals within the scientific community is still not universal, partly because electronic journals are said not to meet the criteria of building a collective knowledge base, communicating information, validating the quality of research, distributing rewards and building scientific communities. Accessing electronic publications over the internet can be unacceptably slow, and incorrect or out-of-date electronic addresses can be a source of considerable frustration. A study by Harter and Kim (1996) revealed that 55% of the links to e-journal archives did not work at first try. Nevertheless, electronic journals have numerous benefits to the user that seem to be outweighing the problems: they are accessible wherever the users need them; they are usually published faster and they can be widely distributed in a network environment.

Online databases

Online databases are large collections of machine-readable data that are maintained by commercial agencies and accessed through communication lines. Many libraries subscribe to online databases because they provide ease of use and current information, and they can be searched through multiple access points. The disadvantage of most online databases is that they do not give full text but only bibliographic data. Like other web resources, online databases also have the problem that information cannot be accessed when the system is down for any reason.

Models of remote reference services

There are various approaches to providing web-based reference services, but the most common are e-mail or interactive video based. The two methods can be complementary, resources permitting. Whatever model the user chooses, the procedure for offering web-based reference service according to Abels (1996) consists of the following communication stages:

- problem statement by the client
- summarization by the intermediary
- confirmation by the client.

Summarization entails a summary of needed information prepared by the librarian, and an outline of the characteristics of the required answer. The client confirmation occurs after the librarian has sent preliminary or final results of the search, and the client indicates whether the information need has been met.

Although e-mail is often used in this process, there are a number of challenges that this presents. When the system undergoes a period of instability, communication breaks down; there is the possibility of missing or incomplete information with users omitting information that is essential to answering a query; and staff can omit essential information in answering queries. The problem of missing information can be resolved in part by extending the back-and-forth dialogue. Other limitations of e-mail communication include lost time between messages and loss of message richness.

However, e-mail communication also has a number of benefits that generally outweigh the problems. There is expanded access to the service; often there is greater convenience for users; and staff experience enhanced efficiency as a result of the asynchronous way in which the medium can be used. E-mail has also the benefits of having a broader base of potential users because of its more widespread use and greater convenience in terms of service hours, again because of its asynchronous nature. E-mail also does not require a significant investment in terms of additional reference personnel, as questions can be handled on an as-available basis, or even distributed to other library personnel. The basic restriction that needs to be built into such service is the maximum time to answer – perhaps less than 48 hours.

The e-mail approach can make use of a web template that might be a simple pop-up form pre-addressed to a reference librarian's e-mail account. At its simplest the form could ask for the e-mail address of the sender, which could be system-supplied, with guidelines about entering the subject data of the request provided. Some libraries may require more complete information beyond simple identification and the reference question, such as asking the user to indicate sources already consulted, keywords and a set of dates.

The other common method for providing web-based reference services is through interactive video. This approach has the advantages of greater media richness and immediacy of interaction because of its synchronous nature. Video-based reference services can be offered at prescribed times, and staff need not be continuously dedicated to the task or tied to a video-equipped workstation for the duration of the service hours (Lessick, Kjaer and Steve, 1997) Video reference services are closer to face-to-face reference services than the e-mail approach. Because of its staff- and technology intensive nature, it is important to limit the number of sites from which such sessions can be initiated. The library might want to identify existing computer laboratories or remote computer centre sites that are equipped for such services, as it is important to build on existing campus and library technical infrastructures to minimize the cost (Sugimoto et al, 1995).

Scheduling video reference sessions can be achieved using a web-based template similar to the form used for e-mail queries. The use of such a form allows the librarian to do some advance preparation by having some personal and subject data in advance. The format of the video reference service may involve videophone service, where the user schedules a session with a librarian, goes to one of the authorized remote sites at the appointed time, and opens the session. The librarian would be waiting, prepared to deal with the specific user and the specific request.

Training needs for web-based reference services

The implementation of the web-based reference services must commence with training of the staff who will provide the service. The training components often consist of small modules, each building on the previous one. Staff need basic computer skills to access the internet – Windows navigation, using icons and menu bars, opening windows and folders, creating, saving, moving files and making internet connections (Metz and Junion-Metz, 1996). More advanced skills include the internet and how it works, what servers and clients are and how they communicate, web browsers and what they do, hypertext links and how they work, home pages and URLs. Retrieval skills would include searching the internet, finding information on specific topics, determining the provenance of web information, knowing how and where to do the general search, understanding search results, where to find graphics, finding and using helper applications.

Staff would also need some basic knowledge of computer networking and telecommunications to appreciate the opportunities and constraints that will affect them. They should also know how to use HTML editor programs, as this will help them to keep the reference website up to date, and they should also be trained to have clear understanding of tools on the internet such as Telnet, Gophers, and File Transfer Protocol. The staff providing web-based reference services need to understand the function of URLs, especially how URLs are embedded into HTML documents to make it possible for one to connect to and create links from one web document to another. They should also know how URLs are used when citing internet information in bibliographies and footnotes.

Users on their part would need to be provided with the skills that they will immediately put to use, such as using browser toolbar options, hypertext links, questions and answers about the resource, tips for using the resource, technical support information, and where and when to get further assistance. The users should have a mix of lectures and hands-on practice using focused exercises, and they should also be taught about the information resources relevant to their needs (Metz and Junion-Metz, 1996).

Planning, design and implementation of web-based resources

The design and implementation of a website starts with defining the purpose and goals of the library in general and the web-based reference services in particular. This has to be followed by defining the collection, user characteristics and their information needs. Decisions have to be made regarding which resources to collect, whether they are computerized bibliographic services, full-text databases, multimedia resources, journal articles, etc. It is important in developing web resources to maximize appeal to a variety of users to enhance usage.

The service provider needs to define the service function of the web resources as well as the means of accessing the service so that necessary infrastructure can be put in place both at the provider's and user's site. The planning of providing web resources should also consider the skills that are needed to access the collection, as well as the speed and type of internet connection that will be available to users. Decisions have to be made on the type of web browser that will be used to access the information, and it is important to ensure that the information display in the chosen browser is clear. The technological capabilities of the users and their receptivity to a new information environment need to be taken into account if the web resources are be utilized optimally.

In designing the web service consideration should be given to the hours when the service can be provided, whether the service will be a 24/7 service or have prescribed hours. It is also necessary to consider who will provide the technical support. Decisions also have to be made regarding the mode of communication with users – whether by e-mail, videophone, etc. Knowledge of network speed must be taken into account, especially if multimedia features are to be integrated into the content. The mode of access to the resource needs to be taken into account, though often this tends to be determined by financial, technological and physical locations.

A web collection development plan needs to be put in place; this might take into account tools such as webliographies and other specialized information services that can be used to identify web-based resources such as meta-sites, e-libraries, directories, subject guides, e-journals, subject gateways, search engines and databases. Once the resources have been identified, it is important to evaluate and select those suitable for the user community. The evaluation of these resources should extend to the finding tools, including sites that review resources.

Content design

Stover and Zink (1996) list the following content considerations in designing a web page:

- a descriptive title
- a header that names the organization or sponsoring body
- a concise paragraph that describes the purpose of the site
- contact information identifying the author
- the date of last revision
- a table of contents as a starting point for reaching important linked documents
- multiple paths for reaching the same information
- links to the home page starting point.

The design has to ensure that the most important information is placed where it can be located easily, preferably on the home page. Generally visitors to a website are known to start reading at the top left of a page, work their way across and down, moving left to right until they get to the bottom right corner. The most important pieces of information on each page are generally found in a headline or in a paragraph of the text. Footer information often contains navigation buttons and links for feedback. The footer also contains text information such as the document's URL, the name of the person responsible for the content, the date the page was last revised, a copyright statement, the library logo, address, phone and fax number.

Usually colours add value and attract visitors to such sites, so they should be used to create visual 'excitement' and encourage the user to read further. Wherever pictures are used, they need to be relevant to the text around them and must be of high quality. In addition, the picture has to be consistent throughout the series of documents. Pictures and graphics in general slow down the loading time for a web page, and may cause the user to seek faster-loading pages (McCready, 1997). The building and updating of a site need to be a daily practice, and the date when the site was last updated should be stated clearly (Garlock and Piontek, 1999; Davidsen, 1997).

Before the website is launched it must be thoroughly tested to ensure that:

- it displays correctly using various types of browser software
- it can be accessed through the library's local-area or wide-area network
- the internal navigation links between pages work correctly, and the external links connect to live sites
- the information provided for users and staff is organized effectively
- the text is easy to read and the graphics are relevant
- important information is easy to locate.

Once the website has been tested and found to meet the relevant benchmarks, it has to be publicized through press releases, radio interviews, a market launch, and perhaps other websites.

Organizational schemes in library web design

Metz and Junion-Metz (1996) identify three basic organizational schemes that can be employed to structure a library website: hierarchical, linear and interconnected. In the hierarchical organization the website is built on a pyramid model as depicted in Figure 8.2, with the starting place the home page containing links to other major sections, each of which has sub-sections. This is the most common form of web organization, and libraries often use it because it mirrors the way the libraries themselves are organized.

The strengths of the hierarchical model of organizing web pages are that:

- it presents users with a starting point that makes their search for internet-based information as logical as using the online catalogue to locate a book in the stacks
- it facilitates ease of access to the information

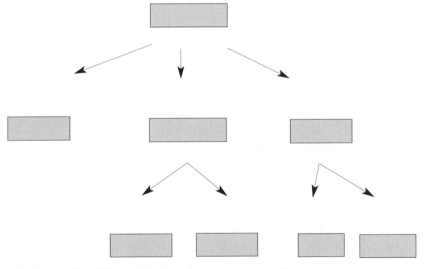

Fig. 8.2 *Example of hierarchical web page organization*
(Metz and Junion-Metz, 1996)

- it is easy to create and maintain because it reflects the way libraries already arrange information.

The problems with the hierarchical structure are that:

- it assumes that all users search from the same starting point
- it also assumes that users always start their search from the home page
- it does not allow easy customization of approaches to finding information.

A second scheme of organization is the linear approach shown in Figure 8.3. This assumes that users start from the web page and then proceed from one page to another in a set progression. This is an easy web structure to organize in the abstract, because it moves logically from page to page. However, it is not easy to create in reality because it covers a considerable amount of information needed by users.

The main advantage of this type of organization is that it can be used effectively within either a hierarchical or interconnected web structure, and it leads users through a series of instructions or steps in a systematic progression. The disadvantage of this structure is that it mandates what users look at rather than giving them a choice. In general users want the freedom to explore information in their own way and react negatively to excessive use of linear organization.

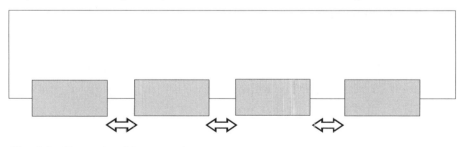

Fig. 8.3 *Example of linear web page organization*
(Metz and Junion-Metz, 1996)

The third form is the interconnected web organization shown in Figure 8.4. This model assumes that there are multiple starting points for exploring the website. Users can move freely through pages and find information based on their own needs, because they are interconnected to every page. Thus every page is a potential home page, because it assumes that users access the web from different points of view.

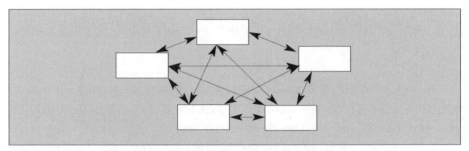

Fig. 8.4 *Example of interconnected web page organization*
(Metz and Junion-Metz, 1996)

This model does not assume that libraries know what information users need. It is a forward-looking type of web structure, as it gives users the feel of how the net and the web are structured. Also, it is relatively simple to create, since every page is linked to every other page. The main disadvantage of this arrangement lies in the fact that it forces users to explore on their own, compelling them to make unaided decisions about what they wish to see.

Future challenges and opportunities

Law and Nicholson (2001) identify several issues that have an impact on the development of effective digital reference service, and all of these need to be resolved as we move into an increasingly digital era.

The first, and perhaps most pressing, of these are concerns about the security and copyright of digital information. At present it is difficult to prevent illegal copying without the use of technologies such as encryption, which in turn make it extremely difficult to access and use information. Systems need to be more secure so that unauthorized access is less likely, and there need to be protocols in place to prevent illegal copying, at the same time as producers and users seek to resolve the present stalemate regarding copyright.

Second, the web is a rich repository of information, but this richness counts for little without more reliable and efficient ways of retrieving and extracting the information. At the input stage the automatic processing of information makes it easy to load data on to the web, and we need equally quick and reliable means of downloading only the required information. We need systems that automatically rank information according to relevance so that bias is less likely in the retrieved information. At the searching stage users need clearer guidelines on how to narrow the focus of their searches without necessarily relying on a human intermediary, and there needs to be more similarity between search engine protocols so that they might retrieve broadly similar information for the same search queries – at present search engines can return remarkably different results for the same searches. A related problem is that the search engines are difficult to use and take too much time to master effectively.

Third, popular perceptions to the contrary, web-based information is never 'free' and in fact can be rather costly for several reasons. For one, web directories are labour intensive to maintain and require regular, detailed attention in order to remain up to date. Further, because the volume of information on the web is enormous and not easy to retrieve effectively, there is a growing need for specialist intermediaries to guide users through the available resources and technologies. Also costly in real terms is the change in library operational practices when digital services are introduced – cataloguing policies may change, staff require retraining, new technologies must be acquired, etc.

Despite these major problems web-based reference service will play an increasing role in the services that libraries provide, and it is important to seek ways of overcoming the problems sooner rather than later. This chapter has sought to show some of the current trends in addressing these problems and other issues related to digital reference service from the information professional's standpoint.

Conclusion

The provision of information services to the user has been enhanced by the web, but not without a wide range of challenges to the information provider. Libraries that are not ready to provide web-based services are becoming increasingly isolated as users are discovering what they can gain from the web and are going where they can access such service if their libraries cannot provide for their cyber needs. Libraries must re-evaluate their mode of information delivery to make it more effective and conform to current trends in order to continue having a place in the information economy, and libraries are realizing that it is either business through the web or they are out of the business. Libraries which are able to exploit the web to provide reference services are benefiting from inexpensive compact storage, faster provision of access to information they hold, and an ability to meet the user's needs, with access to a diverse range of web-based resources.

The challenges facing libraries in their endeavour to use the web to provide reference services include costly investment, inadequate infrastructure, lack of relevant software applications and efficient telecommunications for both the library and the user for effective communication, undeveloped standardization to provide a common interactive interface between the user and the information provider, and the instability of web-based information resources. Even thornier issues – among them global co-ordination, issues of copyright and intellectual property, variety of classification and cataloguing standards, authenticity and authority of web resources – must be borne in mind when planning for web-based service. Library staff must acquire additional skills in order to act as effective intermediaries with virtual users. Such skills are also vital in the identification, evaluation and selection of the diverse web-based information resources.

For modern libraries everywhere in the world the provision of web-based service to the virtual user is a key determinant of their very survival and *raison d'être*, even in the world's less developed countries. Quality, speed, efficiency, cost-effectiveness and meeting specific needs of users are likely to become the virtues and hallmarks that will differentiate one library from the other in the current digital dispensation. Web-based reference services will become more widespread and sophisticated as the web becomes commonplace throughout the world, and to be successful players in the e-world libraries must continue to address the web design and implementation issues raised in this chapter.

References

Abels, E. G. (1996) The e-mail reference interview, *RQ*, **35** (3), 345–58.

ADAM Consortium *The Art, Design, Architecture and Media Information Gateway*
www.adam.ac.uk/adam/index.html
(accessed 20 February 2002)

Argus *The Argus Clearinghouse*
www.clearinghouse.net
(accessed 21 February 2002)

Brown, E. W. and Duda, A. L. (1996) *Electronic publishing programs in science and technology*, available at
www.library.ucsb.edu/istl/96-fall/brown-duda.html
(accessed 10 January 2002)

Burnett, P. and Seuring, C. (2001) Organising access to free internet resources: an overview of selection and management issues in large academic and national libraries with a view to defining a policy at Oxford University, *Program*, **35** (1), 15–31.

Corbin, B. G. and Coletti, D. J. (1995) Digitisation of historical astronomical literature, *Vistas in Astronomy*, **39** (2), 127–35.

Cornell University Library (2001) *Critically analysing information*

resources, available at
www.library.cornell.edu/okuref/reserch/skill26.htm
(accessed 2 May 2002)

Dan, J. K, Welsh, S. and Martin, J. K. (1998) Cross searching subject
gateways: the query routing and forward knowledge approach,
DLib Magazine, (January), available at
http://mirrored.ukoln.ac.uk/lis-journals/dlib/dlib/dlib/
january98/01kirriemuir.html
(accessed 21 February 2002)

Davidsen, S. L. (1997) The Michigan Electronic Library, *Library Hi
Tech*, **15** (3/4), 101–6.

eXTReMe Digital. eXTReMe Tracking
www.extreme-dm.com/tracking/
(accessed 2 May 2002)

Flanagan D. and Lauderdale, F. (2002) *Subject directories*, available at
http://home.sprintmail.com/~debflanagan/subject.html
(accessed 2 May 2002)

Gardner, M. and Pinfield, S. (2001) Database-backed library websites: a case study of the use of PHP and MySQL at the University
of Nottingham, *Program*, **35** (1), 33–42.

Garlock, K. L. and Piontek, S. (1996) *Building the service-based library
web site: a step-by-step guide to design and options*, Chicago: American
Library Association.

Garlock, K. L. and Piontek, S. (1999) *Designing web interfaces to
library services and resources*, Chicago: American Library
Association.

Harter, S. P. and Kim, H. J. (1996) Accessing electronic journals and
other e-publications: an empirical study, *College & Research
Libraries*, **57** (5), 440–56.

Indiana University North West Ask a Librarian
www.lib.iun.indiana.edu/askalibn.htm
(accessed 21 February 2002)

IFLA (1999) *Digital libraries: cataloguing and indexing of electronic
resources*, available at

www.ifla.org/IL/catalog.htm
(accessed 2 May 2002)

Law, D. and Nicholson, D. (2001) Digital Scotland, the relevance of library research and the Glasgow Digital Library Project, *Program*, **35** (1), 1–14.

Lessick, S., Kjaer, K. and Steve, C. (1997) Interactive reference services (IRS) at UC Irvine: expanding reference service beyond the reference desk. Paper presented at the ACRL Conference 1997, available at
www.ala.org/acr/paperhtm/a10.html
(accessed 2 May 2002)

Library of Congress. Cataloguing Policy and Support Office (1997) *Announcement of draft interim guidelines for cataloguing electronic resources*, available at
http://lcweb.loc.gov/catdir/cpso/elec_res.html
(accessed 2 May 2002)

McCready, K. (1997) Designing and redesigning: Marquette Libraries' web site, *Library Hi Tech*, **15** (3/4), 83–9.

Metz, R. E. and Junion-Metz, G. (1996) *Using the world wide web and creating home pages*, New York: Neal-Schuman Publishers.

Morris, S. (2001) The pros and cons of electronic journals, *INASP Newsletter*, **18** (October), 2-3.

National Information Services and Systems (2001) *Directory of networked resources*, available at
www.niss.ac.uk/subject/index.html
(accessed 2 May 2002)

Sloan, B. (1998). Service perspectives for the digital library: remote reference services, *Library Trends*, 47 (1), 117–43.

Stover, M. and Zirk, S. D. (1996) World wide web home page design: patterns and anomalies of higher education library home pages, *Reference Services Review*, **24** (3), 7–20.

Sugimoto, S. et al. (1995) Enhancing usability of network-based library information system: experimental studies of a user interface for OPAC and collaboration tool for library services. Paper

presented at the Second Annual Conference on the Theory and Practice of Digital Libraries, June 11–13, 1995, Austin Texas, USA, available at
http://csdl.tamu.edu/DL95/papers/sugimoto/sugimoto.html
(accessed 2 May 2002)

Traffick.com (2001) *Guide to portals and search engines*, available at www.traffick.com/directory/portals/enterprise.asp
(accessed on 2 May 2002)

University of Bristol. Information Service. Social Science Information Gateway
www.sosig.ac.uk/
(accessed 2 May 2002)

University of Wolverhampton. School of Computing and Information Technology
www.scit.wlv.ac.uk/wwwLib/
(accessed 2 May 2002)

Wusteman, J. (1996) Electronic journal formats, *Program*, **30** (4), 319–43.

Yahoo, Inc. Yahoo
www.yahoo.com
(accessed 21 February 2002)

9

Web-based reference services: the user-intermediary interaction perspective

Sherry Shiuan Su

Introduction

Emerging information technologies are allowing libraries to transform the way of providing access to their resources and, consequently, changing the way users find and utilize information. The digital library offers users the prospect of access to electronic sources at their convenience temporally and spatially. Many libraries are experimenting with the provision of reference services via a variety of technological means and using web-based resources to answer queries. In order to provide equal service to users in and outside the library, libraries have implemented off-site reference desks for responding to requests from off-site users. Librarians answer reference questions via telephone, e-mail and online chat.

Websites and web-based reference services share a similar goal – to provide information and services to a set of users. Therefore, understanding how users seek information on websites and interact with intermediaries is directly relevant to the design of web-based reference services. It is important that knowledge of users' tasks and uncertainties enables us to provide better tools. Intermediaries, as a link between users and information, need to play a crucial role in helping users express their information needs. Negotiation and translation aspects may be critical in getting the user to the needed information.

Much has been written about the web-based reference service, and the focus of most studies has been on the technology or on the general descriptions of the service. Human interaction in web-based reference work is discussed less frequently in the literature, and the purpose of this chapter is to explore the problems and issues concerning the characteristics of web-based reference service from the user–intermediary interaction perspective.

The value of web-based information sources

The internet affords reference services the ability to conduct entire reference transactions in a digital environment, from specifying user needs to delivering information from the collection. The advent of web-based reference services has increased the chances for the successful resolution of many reference questions and the demands these resources place on the librarian's skills (Tenopir and Ennis, 1998). As Dickstein, Greenfield and Rosen (1997) indicate, the internet is proving to be an especially suitable medium for accessing and using reference materials. Griffiths (1999) believes that the web is an additional format to consider when deciding how to meet demand in a subject area. For example, a public library in Pagosa Springs, USA, found that its collection was deficient in medical reference materials. The library decided to rely on web resources to supplement its medical reference materials and listed it as a selection objective in the collection development plan.

Unlike traditional reference tools, the web is not a discrete, uniform quality resource. Librarians are aware that the internet is different from a library. It is: (1) located everywhere and anywhere; (2) disorganized, unstructured, and flat; unverifiable; digital only; and (3) doubling in growth each year (Kresh and Arret, 2000). 'Along with the valuable, accurate, and responsible material on the Internet, there are sources to be questioned, scrutinized, or ignored' (Dickstein, Greenfield and Rosen, 1997). Atkinson (1996) suggests that libraries take responsibility for selecting a subset of resources from the internet. Library users can benefit from the selection and evaluation processes because it saves them time.

Characteristics of web-based reference service

According to White (2001), digital reference service can be defined broadly as 'an information access service in which people ask questions via electronic means (e.g., e-mail or Web forms)'. In this chapter the terms 'digital reference service', 'web-based reference service' and 'electronic reference service' are used interchangeably as terms with similar meanings.

Many articles have focused on augmenting traditional reference services with internet resources and capabilities. Lankes (2000) lists five reasons for moving to electronic reference service:

* increasing access to resources beyond the library
* lack of geographic constraints for users
* the need to differentiate services to different populations of users in the face of shrinking budgets
* increases in complexity of information resources and the need for specialized knowledge
* new options for answering reference questions.

In discussing the characteristics of e-mail reference services, Lagace (1999) emphasizes that e-mail is a time-delayed text interaction between two people who may never have met In a face-to-face reference interview a librarian can ask some auxiliary questions immediately or discern details in other, more subtle ways, but the time delay and remote nature in e-mail communication prevents the give-and-take of a conventional reference interview.

Advantages and disadvantages

As an innovation in information access, web-based reference service is relatively easy to implement. It is also flexible, so it can be tailored to suit the needs of individual libraries and their users. Compared with conventional reference service within libraries, the advantage is that responding to web-based reference questions can be shifted to non-peak load time and to the appropriate respondent, thus optimizing staff resources (White, 2001). Other advantages for a web-based reference service include 24-hour access for users, reduced barriers, easy referral of questions, option to build question/answer databases, and utilization of new technology (Fishman, 1998). As Lagace (1999) concludes:

> reference service on the Internet requires many of the same qualities as traditional reference: accuracy, promptness, courtesy, an understanding of the information need. And while there may be a disadvantage in not having a face-to-face encounter, there are many advantages to this new medium . . . . And the greatest advantage [is that] many more people can be helped by using electronic reference services.

According to Fishman's analysis, librarians should consider several challenges before establishing an electronic reference service (Fishman, 1998). First, questions will take longer to answer, because keying in a response will take more time than answering the question verbally. Besides, the questioner is not physically present during the virtual reference interview, so answers must be more detailed. Furthermore, users cannot easily be referred to a reference book or index to search for answers on their own. Reference e-mail is also very time-consuming because the original query often does not include all the information needed by the librarian. Archer and Cast (1999) also emphasize that conducting an effective reference interview may require several e-mail messages between user and librarian. New 'interpersonal' skills may be needed in an electronic environment.

Second, non-verbal cues are lacking in a virtual reference interview. As a reference librarian, Sarah Haman (2002), says, 'one thing I notice about online interviews is that we can often be misunderstood. Sometimes body language or tone of voice is very helpful.' With an e-mail reference transaction, many librarians assume they understand the question rather than taking the time to clarify the request. Often the exchange may be frustrating for both librarian and user (Archer and Cast, 1999).

Third is the problem of responsibility sharing; answering a question may take days to communicate because the reference staff rotate in most libraries, and staff coming on duty may find it difficult to know the history of the question. Fourth, all responses are available to the user in a potentially permanent and reproducible form. This may seem like an advantage, but special care must be taken in e-mail exchanges to be certain that, if the message is forwarded out of context, it will not be misinterpreted (Fishman, 1998).

In Coffman's opinion one of the most obvious problems of e-mail reference service is that e-mail does not offer the instantaneous response and immediate gratification that users expect from the web (Coffman, 2001). If a question needs clarification, it is also difficult to conduct an effective reference interview using e-mail. E-mail reference is more labour intensive than traditional reference desk service, because it places most of the burden of answering the question on the librarian. A study by Hahn (1998) used content analysis of service logs and interviews with staff and users of a successful e-mail reference service to identify the impacts of the communication medium on service provision. From the research question asked in the study (what do participants perceive as the benefits and limitations of using the medium for the type of service provided?), Hahn was able to determine that both staff and users viewed increased efficiency as the chief institutional benefit and saw improved convenience as the primary user benefit. The limitations include lost time between messages, loss of message richness, and system instability. Hahn concludes that whatever drawbacks or frustrations participants may experience, the frequent use of the service suggests that further development will continue.

Olkowski (2001) points out the pros and cons of the online chat reference. The positive side includes (1) immediate help, (2) internet connection maintained, (3) quicker than e-mail and (4) interactivity. The negative side includes (1) awkward transactions, (2) reference interviews may be more problematic, (3) easier to say it than write it and (4) 'netlag' and congestion.

The virtual reference interview

As Lankes (2000) notes, 'reference services can be defined as mediated interfaces between users in an anomalous state of knowledge and a collection of information'. The user's anomalous state of knowledge is often thought of as a question that needs to be answered. The notion of 'question' and 'answer' is one of the first issues that digital reference must face.

Three elements included in the user–intermediary interaction are as follows: (1) the reference interview, (2) follow-up interaction and (3) delivery of the final product that addresses the client's information needs. Abels and Liebscher (1994) emphasize that the reference process should be viewed as a process that satisfies user needs rather than concentrating on answering the user's question. As a vital part of the reference process, the reference interview also plays a critical role in information systems. White (2001) says that 'the reference interview plays a critical role in information retrieval systems. It is an adaptive mechanism, i.e. it allows for adapting the system to the client or vice versa so that a reasonably congruent match between what the client needs and what the system can identify occurs.' She also claims that a good interview is not sufficient for good results, and the probability of a good end product is enhanced if all relevant factors are identified.

In studying the use of e-mail to provide library reference services to users Abels (1996) discussed the differences between reference interviews via e-mail and reference interviews using other media. E-mail reference interviews were analysed using interview analysis, message counts, subject and motivation, media used, and student critiques. The conclusion of this research was that some substantive reference questions can be negotiated successfully via e-mail. In addition to studying the effectiveness of e-mail reference Abels also used the study to formulate a model e-mail reference interview. The model e-mail reference transaction should consist of three messages: 'the initial form submitted by the user, a summary of the librarian's understanding of the request sent to the user, and confirmation of the summary message by the user' (Abels, 1996).

Without being given any guidance concerning how to conduct a reference interview, each student develops an interview strategy. Abels (1996) contends that five different interview approaches were used by students:

- *Piecemeal* – students asked questions as they occurred to them.
- *Feedback* – the characteristics of the medium allowed students to go back and forth between the various stages of the reference interview process.
- *Bombardment* – some confused or frustrated students would string together a series of different questions in one e-mail message.
- *Assumption* – students would make assumptions about the nature of the client's information need.
- *Systematic* – students would respond in a manner that organized the interview in a structured way when the client sent an unstructured question. Often students would structure their responses by creating something that resembled forms used for face-to-face reference interviews

Conducting an effective reference interview for a complex information problem is a difficult task. There is agreement among reference librarians that a face-to-face reference interview is an important part of the reference process, but the provision of reference services over an electronic network is replacing the face-to-face environment. Abels and Liebscher (1994) point out that many non-verbal cues are not present in an electronic interview, and iconic representations of non-verbal communications are still quite limiting. New and rich protocols for communicating non-verbal signals will no doubt be developed to fill the void.

Straw's analysis (2000) reveals that the face-to-face interview is often enriched by the physical presence of the library collection The librarian and user are able to clarify a request through face-to-face questioning. 'Library buildings are a natural forum for users and reference librarians to solve information problems, and library collections allow for many of these problems to be handled in a single encounter' (Straw, 2000). The electronic interview, however, is not restricted by many conventional boundaries. By using e-mail, librarians and users can communicate from any point in the world. In addition, this approach can allow a more thoughtful process of question negotiation, and more coherence and detail can be given to the possible sources that might be relevant in sorting out a user's particular question (Tibbo, 1995).

In providing live digital reference service to extended campus users, librarians at Temple University in the USA have found that interacting online with a user is quite different from face-to-face and phone transactions. Chat requires a more telegraphic style. Short, to-the-point sentences are preferred over wordier responses. Text-based messages can be a frustrating way to communicate. The time required to construct more elaborate sentences can cause the person on the other end to lose patience. For some of the librarians, the concern that the user might tire of waiting and quit the conversation added anxiety and created a negative aura about the whole service (Stormont, 2001).

Finally, there are two key issues in conducting reference transactions in an internet-only mode; one is scalability, and the other is ambiguity. Originally coined in the manufacturing sector, scalability refers to a process whereby production of an item can be increased to maximize efficiency and effectiveness. When discussing reference services, the question still remains: how can a digital reference service grow, or scale, to handle a large number of questions, given that traditional scaling mechanisms such as service hours and geographical constraints go against user expectations on the internet? The most common digital reference scalability solution in use today is asynchronous reference. It uses either a web page or e-mail to take a question into a service at any time, and to answer it when resources are available. Such a system allows for easy networking of services and scheduling of resources (Lankes, 2000).

Ambiguity is concerned with identifying the resources needed to meet user needs before answering a question. Lankes (2000) suggests that the key to handling ambiguity seems to be providing an appropriate level of interactivity. Therefore, whether real-time or asynchronous, it important for the process to provide an opportunity for users to best express their information needs and ensure that they have every opportunity to seek help, refine their questions, and ultimately have their questions answered.

The user and intermediary of web-based reference services

The virtual users

For those who do not use traditional in-person and telephone services, online reference services can be viewed as a safe, relatively anonymous mode of communication with library staff. Some reference librarians have begun qualitative assessments of those who use virtual reference services and the nature of their questions. Gray (2000) indicates that the people who enjoy working on their own and desire greater equality in the way that users and reference librarians interact are most favoured in the digital world. And the approachability of reference staff online may be higher than meeting face to face. Straw (2000) also comments that the perception of accessibility sets the tone for the user's entire encounter with the library. As users' expectations about the library increase, whatever the method of soliciting questions, the issue of approachability is just as important in the electronic environment as it is at the reference desk.

A reference librarian at the University of Chicago reported that the types of reference questions received online can be separated into two:

- *reference*: data/statistics, finding online journals, searching databases
- *computer problems*: accessing restricted databases, getting on to the network (Olkowski, 2001).

Carter and Janes (2000) found that users seem to have difficulty in assigning subject categories to their questions and determining whether they are factual or require sources for assistance. Fishman (1998) at the University of Maryland Baltimore Health Sciences Library found that approximately 15% of reference questions sent by e-mail were incomplete. In many cases it would be users' experimentation with the system that would result in these partial queries. However, follow-ups conducted by librarians always result in complete questions being returned by the user (Fishman, 1998).

The reference intermediary

The core of reference is mediation. Mediation can be performed either by a human expert or an automated interface. The primary purpose of the interface is to match the user's information need to the system's organization and capabilities. And the intermediary's main function becomes being the user's advocate to the system or collection (Lankes, 2000). Brewer et al. (1996) stress that intermediary services should play a crucial role in the ongoing development of digital libraries. Three major purposes for intermediation in the digital library environment were identified by the authors:

- interaction with potential information beneficiaries
- interaction with information resources
- mediation between information resources and users to add value during the information transfer process.

Sloan (1998) comments that, while the authors do not focus solely on human intermediaries, a very strong case for the necessity of intermediaries has been made by them. And librarians will continue to have a role in the future of networked information. It will involve collaborating with users and information seekers, playing an intermediary role and providing value-added information services, much as librarians do in the physical library setting.

According to Moore (1998), four models for managing electronic information requests can be used by the library:

1 The questions are answered by the librarian who is responsible for the system.
2 The head of reference distributes the questions.
3 The questions go to a department e-mail account and the questions are answered on a first-come first-served basis.
4 Questions are forwarded to the Bulletin Board Service and volunteer librarians answer them on a first-come first-served basis.

As we have already seen, the ideal e-mail reference interview should consist of three messages: statement of the problem by the user, summary by the librarian and confirmation by the user. To keep the interview efficient, Abels (1996) further suggests a systematic approach. It includes:

- designing a structured reference question form to be completed by the user. In addition to describing the request, the user is asked to indicate the type of search wanted and any restrictions that might apply, such as languages, date ranges and age groups.
- if a user sends an unstructured question, it is up to the librarian

to provide structure. Librarians can either respond with a standard request form or with questions written in an organized and logical manner, such as a numbered list.

Meeting the needs of people of diverse backgrounds is now one of the most difficult problems at the reference desk. By constantly working in isolation, reference librarians are not giving their best service. Humphrey (1995) suggests that reference librarians need to construct co-operative, computer-linked consortia to eliminate the need to partially or totally study questions that have already been answered by someone else. This method is called reference annealing, which originated from Neil Larson's concept of 'information annealing'. Larson defined this as the use of a form of computer-assisted indexing, such as hypertext links, by people who wish to combine and fuse together their thoughts about an issue (Humphrey, 1995).

Humphrey (1995) emphasizes that the first step in the process of reference annealing is the creation of a virtual reference collection. This imaginary reference shelf made by combining catalogue records of titles from the real reference collections of many libraries would be the 'place' where reference librarians would go to look up electronic sources for their users' questions.

As Humphrey (1995) further suggests, computer technology allows librarians to insert notes about how to use a particular source and to make hypertext links between one catalogue record and another. Theodor Nelson, the person who originated the term, defined hypertext as 'non-sequential writing – text that branches and allows choices to the reader, and is best read on an interactive screen' (Landow, 1992). The functions of hypertext linking can be summarized as follows:

- It mentally transports the librarian around the shelves of the virtual library in ways that subject descriptors cannot.
- It takes a librarian among the records of a library's directories that young people, in particular, might want to use.

- It makes it easier to help users who fall into more than one socio-cultural group.
- It allows documents on a particular subject to be grouped by point of view (Humphrey, 1995).

Although librarians recognize the value of web-based online information sources, as the sources of online information multiply, one can no longer keep a complete mental catalogue of the likely places where individual pieces of information may be found. In the electronic era new tools will have to be built that can deal comprehensively with a dynamic environment. There have been many debates on who should build the directories to information on the internet (Force, 1994).

Improving user–intermediary interaction

As libraries make more digital resources available on the web, user demand for services over the network will continue to increase. The increasing complexity of the electronic library creates additional opportunities to provide virtual reference services to users. However, the major challenge in the current web environment is also brought about by the substantial increase in the complexity of systems and online services. Johnston and Grusin (1995) argue that virtual reference services provide an important means of helping readers use electronic resources more effectively.

Tools that can help

The human touch in the reference process is important to the success of the reference transaction (Archer and Cast, 1999). Libraries often try to personalize services by using the web to make electronic resources available to their community, as well as to create new ones. However, using technology and making resources available electronically is not the best way to retain the human connection for distant users. The question for librarians is how to develop effective systems for responding to electronic information requests, decide who can be served and to what extent we can assist them, all in the most user-friendly way possible so that they feel 'connected' with the service.

Different technology approaches to reference service have been added over the years. Most recently we have seen the introduction of web-based contact centre software, which is designed for answering questions and providing live, interactive customer service on very high-traffic e-commerce sites. These programs include a wide variety of interactive tools that allow agents to push web pages to clients, help clients through catalogues or databases, collaboratively fill out forms or search screens, and share slide shows and other content online (Coffman, 2001). Many libraries have experimented with this approach, and although it has a good deal of promise as a platform for online reference services, much development work has to be done to adapt it for users' needs. Most of these programs currently use chat to communicate with the user. Coffman (2001) thinks that chat is a rudimentary and cumbersome way to convey messages, much less the complex content of many reference interactions. VoIP (Voice over Internet Protocol) technology, which allows the librarian and the user to browse the web and talk back and forth at the same time, may be a solution to the above problem.

Archer and Cast (1999) introduce desktop conferencing technology to go where the questions are and still maintain the human element of reference service. It is an interesting approach to providing real-time reference because of the audio and video capabilities. McGeachin (1999) also agrees that desktop videoconferencing with personal computers and remote application sharing software can be used to provide personal and effective distant reference services to library users. These technologies include the element of visual contact that can be crucial for conducting a reference interview that elicits the true information needs of the user.

However, according to project results at University of Michigan's Shapiro Undergraduate Library, the reference services between the library and residence halls using desktop videoconferencing technologies have revealed relatively low levels of use. Four reasons have been offered for the lack of success of the project. First is the inconsistent quality of the audio and video connections; researchers found that there was insufficient bandwidth in the implementation to provide necessary levels of service. Second, there was lack of adequate technical support for the CU-SeeMe shareware product. Third, librarians noted that it would have been helpful for librarians and users to be able to see each other's screen displays – 'being unable to see and point to the screen . . . made it difficult to effectively teach the students how to use the library's resources'. Fourth, it was also noted that some students and staff members reported being self-conscious while on camera (Folger, 1997).

In a paper describing an experimental real-time reference service provided in a Multi-User Object-Oriented environment accessible on the internet (Shaw, 1996), participating librarians and users have some suggestions regarding the tools used to deliver a more effective online reference service:

- Simplify all interactions.
- Ensure that procedures are easy to learn.
- Make it possible for on-duty librarians to work at other tasks while waiting for users.
- Use a team approach to answering reference questions.
- Allow the librarian and user to examine material found on the internet at the same time, as this helps user and librarian refine the search.

User satisfaction

The influence of the search interview on the patron's overall satisfaction with online services cannot be over-emphasized. User satisfaction is very hard to define and measure, but Lamprecht (1987) states that it could be summed up briefly as a state of mind experienced or not experienced by the user. According to Hilchey and Hurych (1985), four factors may determine user satisfaction with online searching: the output, the interaction with the intermediary, the service policies, and the library as a whole. Auster and Lawton (1984) designed a series of controlled experiments in order to determine user satisfaction by monitoring the reference interview. The result suggests that a close relationship exists between the types of questions asked in the pre-search interview and the knowledge gain and the satisfaction with the output of search. In the electronic environment, more studies will be needed to investigate users' satisfaction level with web-based reference services.

Important skills needed

The literature shows that there are significant impacts on reference services resulting from greater access to internet tools. These impacts include new skills needed by intermediaries. According to Abels and Liebscher (1994), the success of the reference interview will vary depending on the reference intermediary, the client, the topic and the nature of the query. Three factors will affect the outcome of the reference interview:

- *interpersonal skills* – friendliness, eagerness, kindness and approachability are positive traits for the reference intermediary
- *subject expertise* – knowledgeable intermediaries can provide guidance in the clarification of the query, in selecting the best possible keywords for searching, and in proposing the best reference sources
- *prior experience* – experiences make intermediaries more adept at ensuring that all important points are covered, and that the important points are appropriately synthesized and summarized.

Serving remote users effectively is dependent on the communication ability of all the people involved. To help understand the needs of remote users more effectively, Guenther (2001) raises two useful points. First, interaction with remote individuals allows the intermediary to develop more effective skills and gather specific information about their needs. Second, information, trends, and conclusions can be drawn from statistics that are created as remote users browse the websites. This method is able to provide solid information.

The desirable searcher characteristics suggested by Lamprecht (1987) still apply in today's electronic environment: (1) self-confidence; (2) good communication skills; (3) patience and perseverance; (4) a logical and flexible approach to problem solving; (5) a good memory for details; (6) typing skills; (7) a good knowledge of the subject areas being searched; (8) good organization and efficient work habits; (9) the motivation for acquiring additional information and (10) a willingness to share knowledge with others.

Straw (2000) emphasizes that librarians require certain skills and attributes in order to interact effectively with users online. First, they have to be familiar with the prevailing technology, including its capabilities and limitations, in their library. Second, the important skills needed for being a good listener must be carried over from the traditional face-to-face interview. Third, good written communication is still the key element in making electronic reference service work. Reference librarians should be able to write messages that are organized, concise and logical. 'A well-written response not only answers a question eloquently, but it also tells the user about the importance that the library places on the question' (Straw, 2000).

All in all, as Herbert White (1992) says, successful reference librarians must have wide-ranging knowledge and authentic intellectual curiosity, and intimate and specific knowledge of the tools and resources available. Most important, they must possess the human relation resources to deal 'closely' with remote users.

Quality standards

It is difficult to establish quantitative standards for reference service.

Therefore, reference standards are mostly general and descriptive. Mendelsohn (1997) identified four dimensions of quality reference service: knowledge, willingness, action and assessment. The importance of creating a partnership between the librarian and user in assisting the user in meeting the information need was also indicated.

Kasowitz, Bennett and Lankes (2000) identify a working set of standards to assess individual web-based reference services. For user transaction standards, 'clear response policy' states that clear communication should occur either before or at the start of every digital reference transaction in order to reduce opportunities for user confusion and inappropriate inquires. The goal is as follows:

> State question-answering procedures and services clearly in an accessible place on the service's Web site or in an acknowledgment message to the user. The statement should indicate question scope, types of answers provided, and expected turnaround time.

In the interactive category, it states: 'Digital reference services should provide opportunities for an effective reference interview, so that users can communicate necessary information to experts and to clarify vague user questions'.

The optimal level of service includes the following three goals:

- Offer real-time reference interviews or very thorough web forms to gather as much information as possible without compromising user privacy.
- Allow users the ability to return to a service for further information to clarify a question if the answer is insufficient in terms of the policy guidelines of each service.
- Link related question–answer sets using a common protocol to identify related messages to facilitate follow-up.

Evaluating the quality of the responses to questions is another important component of web-based reference service. Lankes and Kasowitz

(1998) suggest several methods of measuring staff performance, including user surveys, unobtrusive testing, peer assessment, grading or tracking, and self-assessment. At University of California – Irvine, the library staff agree that it is difficult to determine an objective means of evaluating the quality of an answer. Frequently, the reviewer's expertise in the subject or the interpretation of the question becomes the basis of the perception. One of the most effective ways of evaluating the quality of responses is not to leave it to an individual reviewer but to discuss each response as a group (Horn and Kjaer, 2000).

Conclusion

As Lankes (2000) says, 'digital reference is not simply traditional reference work without a desk'. With all of the advantages of taking reference online, there remain many problems with the technology currently available. New technologies will continue to develop that can be adapted to the unchanging goal of meeting user information needs regardless of location or format and in the most convenient way possible (Ciccone, 2001). However, it is clear that digital reference is more than a simple set of techniques and technologies. It is an attempt by the library to come to terms with human intermediation in today's digital libraries. Despite changes in communication technology, the reference interview will remain at the heart of the reference transaction (Straw, 2000). With all the challenges discovered when interviewing users in an electronic age, the ability to understand, question and communicate remains as important as in face-to-face encounters.

References

Abels, E. G. (1996) The e-mail reference interview, *RQ*, **35** (3), 345–58.

Abels, E. G. and Liebscher, P. (1994) A new challenge for intermediary–client communication: the electronic network, *Reference Librarian*, **41/42**, 185–96.

Archer, S. B. and Cast, M. (1999) "Going where the questions are": using media to maintain personalized contact in reference service in medium-sized academic libraries, *Reference Librarian*, **65**, 39–50.

Atkinson, R. (1996) Library functions, scholarly communication, and the foundation of the digital library: laying claim to the control zone, *Library Quarterly*, **66** (3), 239–65.

Auster, E. and Lawton, S. B. (1984) Search interview techniques and information gain as antecedents of user satisfaction with online bibliographic retrieval, *Journal of the American Society for Information Science*, **35**, 90–103.

Brewer, A. et al. (1996) The role of intermediary services in emerging virtual libraries. In *Proceedings of the 1st ACM International Conference on Digital Libraries*, New York: Association for Computing Machinery, 29–35.

Carter, D. S. and Janes, J. (2000) Unobtrusive data analysis of digital reference questions and service at the Internet Public Library: an exploratory study, *Library Trends*, **49** (2), 251–65.

Ciccone, K. (2001) Guest editorial: virtual reference, today and tomorrow, *Information Technology and Libraries*, **20** (3), 120–1.

Coffman, S. (2001) Distance education and virtual reference: where are we headed?, *Computers in Libraries*, **21** (4), 20–5.

Dickstein, R., Greenfield, L. and Rosen, J. (1997) Using the world wide web at the reference desk, *Computers in Libraries*, **17** (8), 61–5.

Fishman, D. L. (1998) Managing the virtual reference desk: how to plan an effective reference e-mail system, *Medical Reference Services Quarterly*, **17** (1), 1-10.

Folger, K. M. (1997) The virtual librarian: using desktop video conferencing to provide interactive reference assistance, available at www.ala.org/acrl/paperhtm/a09.html
(accessed 2 February 2002)

Force, R. (1994) Planning online reference services for the 90s,

Reference Librarian, **43**, 112–13.

Gray, S. M. (2000) Reference services: directions and agendas, *Reference and User Services Quarterly*, **39** (4), 365–75.

Griffiths, J. M. (1999) Integrating the web into reference services, *Colorado Libraries*, (Spring), 10-14.

Guenther, K. (2001) Know thy remote users, *Computers in Libraries*, (April), 52–4.

Hahn, K. (1998) *An investigation of an e-mail-based help service.* CLIS Technical Reports, 97-03, College of Library and Information Services, University of Maryland, available at www.clis.umd.edu/research/reports/tr97/03/9703.html (accessed 3 February 2002)

Haman, S. (2002) Virtual interviews. Discussion of digital reference services, 1 February dig_ref@listserv.syr.edu

Hilchey, S. E. and Hurych, J. M. (1985) User satisfaction or user acceptance? Statistical evaluation of an online reference service, *RQ*, **24** (4), 452–9.

Horn, J. and Kjaer, K. (2000) Evaluating the 'Ask a Question' Service at the University of California, Irvine. In Lankes, R. D., Collins III, J. W. and Kasowitz, A. S. (eds), *Digital reference service in the new millennium: planning, management, and evaluation*, New York: Neal-Schuman Publishers,135–52.

Humphrey, N. (1995) Reference annealing: let's stop reinventing the answer, *RQ*, **34** (4), 459–63.

Johnston, P. and Grusin, A. (1995) Personal service in an impersonal world: throwing life preservers to those drowning in an ocean of information, *Georgia Librarian*, **32**, 45–9.

Kasowitz, A., Bennett, B. and Lankes. R. D. (2000) Quality standards for digital reference consortia, *Reference and User Services Quarterly*, **39** (4), 355–63.

Kresh, D. and Arret, L. (2000) Collaborative digital reference service: update on LC initiative. In Lankes, R. D., Collins III, J. W. and Kasowitz, A. S. (eds), *Digital reference service in the new millen-*

nium: planning, management, and evaluation, New York: Neal-Schuman Publishers, 61–8.

Lagace, N. (1999) Establishing online reference services. In Janes, J. et al. (eds), *The Internet Public Library handbook*, New York: Neal-Schuman Publishers.

Lamprecht, S. J. (1987) Online searching and the patron: some communication challenges, *Reference Librarian*, **16**, 177–84.

Landow, G. P. (1992) *Hypertext: The convergence of contemporary critical theory and technology*, Baltimore: Johns Hopkins University Press.

Lankes, R. D. (2000) The foundations of digital reference. In Lankes, R. D., Collins III, J. W. and Kasowitz, A. S. (eds), *Digital reference service in the new millennium: planning, management, and evaluation*, New York: Neal-Schuman Publishers, 1–10.

Lankes, R. D. and Kasowitz, A. (1998) Virtual Reference Desk AskA Software: decision points and scenarios. White Paper for the Virtual Reference Desk, ERIC Clearinghouse on Information and Technology, Syracuse, NY, ED417728.

McGeachin, R. B. (1999) Videoconferencing and remote application sharing for distant reference service, *Reference Librarian*, **65**, 51–60.

Mendelsohn, J. (1997) Perspectives on quality of reference service in an academic library: a qualitative study, *RQ*, **36** (4), 544–57.

Moore, A. (1998) As I sit studying: web-based reference services, *Internet Reference Services Quarterly*, **3** (1), 30–1.

Olkowski, S. (2001) Digital reference in academic libraries. In *Third Annual VRD Conference: setting standards and making it real, November 12-13, 2001, Orlando, FL, USA*, available at www.vrd.org/conferences/vrd2001/proceedings/olkowski.shtm (accessed 2 February 2002)

Shaw, E. (1996) Real-time reference in a MOO: promise and problems, available at www-personal.si.umich.edu/~ejshaw/research2.html (accessed 16 February 2002)

Sloan, B. (1998) Service perspectives for the digital library remote reference services, *Library Trends*, **47** (1), 117–43.

Stormont, S. (2001) Going where the users are: live digital reference, *Information Technology and Libraries*, **20** (3), 129–34.

Straw, J. E. (2000) A virtual understanding: the reference interview and question negotiation in the digital age, *Reference and User Services Quarterly*, **39** (4), 376–9.

Tenopir, C. and Ennis, L. (1998) The impact of digital reference on librarians and library users, *Online*, **22** (6), 84–8.

Tibbo, H. R. (1995) Interviewing techniques for remote reference: electronic versus traditional environments, *The American Archivist*, **58**, 294–310.

White, H. S. (1992) The reference librarian as information intermediary: the correct approach is the one that today's client needs today, *Reference Librarian*, **37**, 23–35.

White, M. D. (2001) Diffusion of an innovation: digital reference service in Carnegie Foundation Master's (Comprehensive) academic institution libraries, *Journal of Academic Librarianship*, **27** (3), 173–87.

10

Digital library initiatives for academic teaching and learning: towards a managed information environment for online learning

Judith Clark

Introduction

Perhaps the biggest challenge facing academic libraries in the 21st century is the integration of their services into learning and teaching (Brophy, 2001). Although many universities are purchasing learning management systems, their web-based learning resources often consist of little more than electronically distributed versions of traditional learning materials such as course outlines, lecture notes and reading lists. While there are some good reasons for using the internet to distribute traditional materials, this chapter takes the approach that the new computing and networking technologies offer distinctly new opportunities for libraries to enhance the learning process. This requires a focus on systems that facilitate a ready flow into the learning environment of a diversity of types of information. It also requires a closer focus on what information is used by learners and teachers, and the most effective ways of interacting with information in different learning contexts.

Integration of academic library services was the subject of a six-month investigation carried out by INSPIRAL (INveStigating Portals for Information Resources And Learning) during 2001 (Currier, 2001). This UK project was funded by the Joint Information Systems Committee (JISC) to identify and analyse institutional and user issues with regard to linking online learning environments and digital libraries. The INSPIRAL study was a useful review of UK experiences with provision of library services in the context of online learning, but it approached the issues conservatively and on the whole from a traditional library-centric perspective.

This chapter looks at the broader issue of information use in the curriculum, specifically considering how digital library technologies can contribute towards more effective delivery of information services for lifelong learners. It explores what it might mean to create and facilitate an integrated online environment embracing both learning and information technologies.

What possible new applications are emerging from the digital library field, and what tools do these technologies offer to foster higher-level learning outcomes? One significant strand of work in this area is that being led in Australia by McLean (2001), focusing on the interoperability issues underlying the convergence of online learning and information environments. While the technical challenges are significant, McLean feels the major obstacles to achieving interoperability are political. Hawkins (2000) suggests the biggest challenges in meeting the information and support needs of learners in a distributed learning environment are legal and business related. Casting further light on what these dangers might be, Blackall (2002) discusses the difficulties librarians face in seeking to implement university-based information literacy programmes that encroach on academically controlled areas of curriculum design and teaching.

A multidimensional approach is required in rethinking information services for online learning. The INSPIRAL study revealed a variety of initiatives adopted by UK libraries in response to new course delivery technologies (Currier, 2001). Some of these extend existing library services, others indicate the emergence of new roles and processes. How individual libraries respond reflects the strategic goals of the particular institution, and the reasons for any one university deciding to adopt technology-supported course delivery will vary enormously. While the dynamics differ from institution to institution, it is possible to identify some common drivers. The next section considers some of the broader forces driving academic libraries to explore new service models

Factors of change in academic libraries
New modes of educational delivery

Flexible learning is freeing up the place, time, methods and pace of learning and teaching. As more universities implement learning management systems, whether as a means of complementing face-to-face teaching or for full delivery, students are coming to appreciate the benefits of 24-hour access to course materials, and in some cases even degree programmes. These students require library and information services to be delivered flexibly. Library support for online learning can be seen to have evolved out of experiences of providing services for distance education. More advanced support services utilize the capabilities of the new instructional technology in innovative ways. For example, support staff at Edge Hill College in the UK are using WebCT delivery software to scaffold students' development of generic skills, progressively moving to more advanced skills as the students gain confidence in exploring and using information (Roberts, 2001). The Virtual Training Suite (www.vts.rdn.ac.uk/) developed by the Resource Discovery Network is another successful model. The 40 subject-based tutorials each offer an interactive 'tour' that aims to build key information skills for the internet environment. They can be used to help lecturers seeking to incorporate information skills teaching into their curriculum, librarians who need tools to support their user-education programmes, and learners wanting a self-paced resource to help them improve their internet information skills.

New modes of scholarly communication

Arms (2000) points to the declining importance of university libraries in relation to other information sources. The context for his observation is the changing nature of scholarly communication, including the increasing ability of researchers to access electronic publications independently and to exchange large amounts of information from their own desktops. There is a raft of new information delivery services targeting the higher education market, reinforcing a perception common among younger students that the internet is the library. Searching from the desktop is an effective means of accessing the resources they need to support their studies. The twin spectres of, on the one hand, the invisible library and, on the other, the deserted library may be cause for revising service priorities. The invisible library describes a picture presented by Brown and Duguid (2000), where all the information an academic might need is available from the desktop but with no acknowledgement of the chain of professional work that makes that information available. 'The deserted library' is the title of an article in the *Chronicle of Higher Education* (Carlson, 2001) that discusses the introduction of coffee shops as one means of luring users back and revitalizing library buildings. It is with a sense of the threat of readily available alternatives that some librarians are seeking to improve their visibility by integrating their services into the online curriculum.

Technology developments

The potential to use technologies to improve access to and use of their print and electronic collections is for some libraries a driving force for integration. In the digital library arena, research is steadily advancing understanding of what is possible in electronic information delivery. This represents an entirely new strand in the evolution of libraries and is likely to profoundly influence the evolution not just of library services but of online learning and information environments as well. A number of university libraries are developing institutional subject portals for information, and several national projects are developing 'scholars' portals'. These call on technological models that have emerged from digital library initiatives, from the corporate knowledge management field and from the online government agenda. Information portals seek to integrate diverse information resources and tailor them to meet the needs of users. The technology is capability building; as students navigate a learning environment, a decision trail can be built up. By gathering and analysing data on actual use of systems it is possible to dynamically reshape services in response to those data. This type of technology will influence a whole range of the services delivered by universities, but research in the digital library field suggests the most significant impact will be in terms of the emergence of learning communities (Clark, 2002). In the UK the Distributed National Electronic Resource (DNER) agenda is leading a development programme to encourage research and innovation in these areas at the institutional level (JISC, 2002).

A parallel strand of technological research is coming out of the learning technology domain. An online course or subject can become a 'portal for learning'. Portals for learning call on resources relevant to a particular learning programme. Learning Object Metadata (LOM) is a set of guidelines that define small 'objects' as the basis for combining and sharing resources. A specific learning resource can be built up from a number of objects gathered from a variety of digital repositories. A portal environment could be a means of offering learners greater choices of assignments and resources, while providing multiple opportunities for them to call on the assistance provided through library support services.

Librarians need to consider information provision as being initiated by both pull and push requirements. As libraries experiment with new functionalities for targeted (push) services, they are developing new processes for the dynamic delivery of information. The technology offers an unprecedented scalability in delivering customized delivery services. The digital environment also offers new opportunities for learners to interact with information resources, giving them, for example, the freedom to take their own time, to undertake learning tasks that involve identifying and selecting items and interpreting them, and to use them in learning assignments. As it becomes increasingly possible to provide instantaneous access to the actual item from metadata surrogates, it will be easier to use library resources in learning assignments. Increased demand for services then becomes a driver in its own right.

Pedagogy

Some researchers suggest that the essential impact of student interaction with online learning environments is emerging learner control over individual learning (Collis and Moonen, 2001). The technology enables learner choice in type of learning activity, course resources used, level of interactivity and media support. Flexible learning represents substantive curriculum change, not merely change in teaching or delivery. Resource-based learning and the shift from teacher-centred to learner-centred education requires that libraries rethink their role. To weave usage of information resources effectively into the curriculum and culture of the university requires improved understanding of learning theory. What are the essential characteristics of the library environment in terms of stimulating learning? Can these be replicated online? What is meant by a learner-centred approach to information delivery? Knowledge of learning and teaching theory has not been ranked highly as a desirable characteristic of library directors (Hernon, Powell and Young, 2002), but for libraries to support technology-enabled course delivery effectively, there is a growing demand for support staff who have a sound understanding of both learning pedagogy and information science.

Libraries provide a social environment for learners to interact with resources. They have always facilitated peer-to-peer learning activities by providing opportunities for groups of learners to interact (across discipline boundaries), and provide the stimulus for learners to engage with a variety of resources. In a sense it is natural that librarians would seek to create web-based environments for learning that are more like a library than a classroom. Such an environment provides for interaction with a wider world of ideas, beyond one specific content area. It also seeks to nurture learners in a more holistic sense than is typical of a specific curriculum-based learning environment and has an over-riding aim of engendering a sense of belonging to a wider academic community.

In response to changing models of educational delivery, libraries are also developing systems for delivering information when it is needed, so that it can be put to use immediately in the learning context. This is generally achieved by working with teachers to design learning materials that use information sources in a structured fashion to support specific learning objectives.

Consumerism and service quality

With governments encouraging stronger competition between universities and the increased social focus on consumer choice, universities everywhere are making an effort to improve the perceived quality of their services. A common challenge across administrative, support and learning areas is to design systems that can be more responsive to student needs. A web-based infrastructure is emerging in some institutions that allows systems to become more data driven. For the library, integrated systems will mean that data from different parts of the university can be gathered and analysed in new ways. This includes data gathered through a learning management system, which is a new source in most institutions, with potential to complement the data generated by student administration systems. Librarians seeking to give students a more personalized service and greater choice have access to better information about users and their activities that can be applied to system improvement.

Learners who have grown up in a digital world expect to participate in active learning situations, blend learning activities with other social activities, and draw on diverse sources of information (Barone, 2001). These learners will judge the quality of the library and information they receive against a new set of expectations, developed through their experience as consumers in the commercial context.

The quality agenda impinges on libraries in a further way. Increasingly, university cost centres are being required to justify themselves in terms of performance against the strategic learning goals of their organizations. In this context it is critical that libraries demonstrate their value. Integrating the library's resources and services into the student learning environment is likely to be an effective means of increasing return on investments in collections as well as developing students' capability for independent learning. The challenge of how to measure this impact has been taken up by the Association of Research Libraries in the USA as part of the Higher Education Outcomes Research Review programme which investigates strategies for assessing the library's value to the community and explores the library's impact on learning, teaching, and research (Smith, 2000).

A final note on drivers

Relevance and responsiveness of library support services (reference, assistance, training) depend on anticipating user needs and building systems to address those, or reacting to requests from users. Reactive strategies require that users can access services easily and are willing to ask for help. The potential for adapting digital library tools to these ends is being explored and tested. Online learning environments afford many new opportunities to expose the library at point of need. For example, at a particular point in the learning path an action by a learner may trigger the appearance of a specific library message. Interactive technologies offer many ways to help learners feel more comfortable about requesting assistance and more enabled to do so.

At the same time digital library initiatives suggest new ways of using technology to underpin pro-active strategies to meet emerging and new needs of learners. Despite limited understanding of learners' needs in this new online environment, there are many examples of projects and programmes developing information delivery systems to address user needs in specific ways. The 'Anorak' rating developed by the Learning Technology Support Service (LTSS) at the University of Bristol is a diagnostic test that indicates a level of information technology experience against a one to five rating. Once learners have rated themselves, they are encouraged to explore LTSS resources that have been tagged with the appropriate rating. The courses offered by the LTSS for staff at the University of Bristol are promoted as a means of 'upping your Anorak rating' (University of Bristol, 2002). Each library service needs to strike its own balance between developing both reactive and pro-active strategies for flexible delivery, remaining open to the need to change the focus of services as the changing needs of learners are better identified and understood.

Libraries as intermediaries

Libraries have an enduring role to assist learners of all ages in making connections between print and electronic materials and helping them navigate through the vast realms of electronic information available at their fingertips. The INSPIRAL study revealed librarians to be more concerned about integration than are lecturers or the learners themselves (in this case university students).

Cohen (2001) notes that the alliances that higher education learning management systems are forming with other technology and content companies do not include libraries or their vendors. Universities are purchasing course management software that does not even facilitate access to their local library collection. Blackboard, for example, provides a resource centre, which largely points to free journals and websites, and links to commercial sites that provide information for a fee. In a world where the impact of disintermediation is in a sense more real, what is the future for the intermediary role of the library?

The high-level view of how these systems will deliver 'learning materials' to enrolled students assumes that student use of materials is akin to a classroom activity, not a library-based activity. That is, students will interact primarily with the materials provided by the lecturer or tutor, and they will be guided by the lecturer towards additional materials selected for them. Libraries have always supported this type of material use, usually through the provision of 'closed' high-use collections – the reserve book room, for example. Reserve collections provide a small, selected subset of materials, and the way that students interact with those materials is largely prescribed by the lecturer. The distance education tradition reinforced this view by assuming minimal access to library services. Students were provided with course packs to this purpose (Beagle, 2000). It is common for those responsible for the implementation of course delivery software in higher education institutions to use a vocabulary that indicates they view the learning environment as highly controlled and structured. Senior university personnel often perceive learning management systems as 'streamlining the delivery of education'. For example, the 1999 White Paper *Learning to succeed* spoke of 'delivering substantial savings of at least £50 million by cutting wasteful bureaucracy and streamlining programme delivery' (Great Britain. DfEE, 1999) Streamlined delivery, in an administrative context, implies that a relatively narrow range of materials will be made available to the student.

When learning management companies refer to integration, it is usually in the context of integrating data from university enterprise applications such as student administration and human resource systems. There is little or no discussion of the need to integrate with library systems, because, as Beagle (2000) proposes, the old reductionist paradigm holds sway, with its assumption that learning delivery relies on primarily pre-packaged, self-contained resources.

Libraries, in contrast, operate on expansionist assumptions. Libraries have sought to provide a depth and breadth of alternative materials and have sought to make it easy for users to conduct their own searches and to browse collections for items that present different perspectives or formats.

It is possible that access to increasing amounts of information can have a negative impact on learning achievements by overwhelming students. This has been particularly evident in distance learning with learners who already feel isolated and disconnected. Educators refer to instructional materials that incorporate a lot of content as 'shovelware'. Copiously copying texts or other content from some other medium and posting it to the web as course material is recognized as a particularly ineffective way to encourage learning. There are many academics who themselves feel uncomfortable about the sheer volume of information available via the internet and lack strategies for managing the access of their students to digital information. Some are simply reluctant to encourage students to explore because of time pressures. For many reasons there is a perception that instructional materials should avoid giving too much content.

Many of the librarians consulted by the INSPIRAL team spoke of their fears that online delivery would lead to an increase in 'spoon-feeding' students (Currier, 2001). To generalize, librarians are virtuous about the need to integrate into online course delivery the wider range of information resources which serve to broaden the students' knowledge of a subject. Libraries seek to deliver services that reach the highest value, and administrators want to deliver education at the best possible cost. Both assumptions need to be questioned against the potential of new technologies to creative active learning experiences that involve the use and re-use of information resources for coherent learning outcomes.

Librarians need to demonstrate their virtuosity by addressing the valid concerns of senior management and teachers, while determining ways to direct students to their own print collections, electronic resources and highly skilled professional staff. The best way to improve the return on investments in libraries is to ensure that their systems can be integrated into the learning management system.

Libraries have three traditional areas of strength. The first is in the organization of sources and the application of systems and structures. The second is the ability to offer unbiased channels of information, independent of the commercial interests of particular publishers or providers. The third is a commitment to supporting learners, underpinned by equality and equity principles. Course management software that can operate together with library service networks will facilitate delivery of information that is valid and appropriate and allow the specialized training, help and assistance provided by librarians to be embedded where and as needed. The digital library agenda offers the potential to manage the flow of information so that learners are neither spoon-fed nor in danger of drowning at the fire hose.

Organizing sources

Libraries provide management structures and procedures to handle the economic and legal aspects of information provision. Improving the management of information for online learning is about architecture, infrastructure and delivery frameworks. A starting point could be to identify what information and sources need to be delivered. Typically this would include:

- both online and physical information resources
- collections, items within collections, and fragments of information within those items
- both commercial and free information resources
- both locally and remotely held materials
- resources created internally as part of the process of research, learning and teaching
- dynamic community-based dialogue and exchange.

Such an approach may leave little reason to distinguish between what might traditionally be library resources (unguided reading) and what might be called learning resources or course materials (pedagogically guided). Technically it is possible to distinguish between objects that necessarily are of an educational nature (Reusable Learning Objects, or 'RLOs') and objects that simply may be used for educational purposes (Reusable Information Objects, or 'RIOs'). An RLO would contain both learning objectives and prerequisites, whereas an RIO would not. To date, academic libraries have not been greatly concerned with providing access to learning objects. Increasingly, however, and often by default, libraries are taking on new roles in content development, for example, facilitating common metadata standards and content structuring standards, and co-operating with faculty to create digital collections. They are being asked to expand their copyright services, to include advice for faculty on managing the rights aspects of material created in house. Many Australian university libraries are developing rights management databases, and some are incorporating new copyright management requirements into existing library catalogues. The Queensland University of Technology Library has met requirements for e-reserve collections by providing a digitizing service and developing a Course Materials Database (Selby and Young, 2002).

Digital resource development is emerging as a challenge across the university. The aspect that tends to involve librarians is providing access to electronic content. This includes activities to ensure that resources provided for teaching and research are able to be discovered, are high in quality and are kept accessible over time. Libraries provide management structures to ensure that the resources available are used to provide the maximum possible value for money as determined against the strategic goals of the organization. Librarians select and recommend learning resources from a variety of sources, both commercial and non-commercial. In the networked learning environment they are also designing and managing tools that make it easier for learners to access high-quality sources that are freely available on the web. It remains the role of the library to address questions of selection and evaluation of resources and their preservation. Presenting a coherent 'electronic information landscape' which represents an increasingly heterogeneous selection of resources is part of the new challenge (Brophy, Craven and Fisher, 1998).

Accessibility has many dimensions. Dempsey (2000) says university libraries are but one contributor to a 'network space' that is shared with other types of libraries, and also with museums, galleries, archives and government domains. Access to the collections and materials held by a range of different cultural sectors is increasingly achievable (Hawkins, 2000). Learners need not be constrained by institutional boundaries in their discovery and use of information resources.

Infrastructure issues include ensuring that the needs of all members of the university community have equitable access in terms of their ability to use the technology. In 'Growing up digital' Brown (2000) suggests that the web is a transformative medium for social practices. For libraries, new technologies will open up new ways to support skills development to ensure that every member of the community has the basic competencies required for online learning and teaching.

Library users as constituents

It is possible that, as online learning environments mature, certain library services will devolve from the central model to one of discipline- or faculty-level support. Appleton (1999) notes that at Central Queensland University, where the majority of students study in flexible mode, the role of the librarian is more like that of the academic, 'more discipline-based and instructional'. In an environment where information is being drawn from distributed sites and delivered in many formats, a constituency-based model may offer a way for libraries to reorganize around a user-centred framework. Kenyon College in the USA is an example where services have converged around a constituency model (Barth and Cottrell, 2002). In effect, university libraries are being forced to challenge notions of who their users are. Brophy and MacDougall (2000) suggest there is little room for librarians to be complacent as universities position themselves to compete in the lifelong learning marketplace.

Supporting learners

Support of student learning is typically the business of a range of organizational units within the university. It may be useful to think of these as services that optimize learning outcomes by providing opportunities that directly improve the learning experience and services that seek to create an environment in which learning can occur.

One of the aims of the subject resource guides or 'pathfinders' that librarians create is to direct users towards an extended base of information. Through exposure to a wider spectrum of resources, libraries stimulate users to reflect critically on the subject content provided by their lecturers. The traditional subject guide lends itself to being transformed into a web-based, one-stop shop, drawing together the library links and bibliographic databases and other electronic resources in the subject librarian's toolbox. At Canterbury University in New Zealand the library provides a series of subject portals that seek to provide users with a coherent view of the complexities of the resources available (http://library.canterbury.ac.nz/subjects/subjects.shtml).

Subject librarians can complement such guides by forming partnerships with lecturers at curriculum level. Working as part of a course design team, a librarian might include links from the course materials to resources created with other tools. These links might be embedded at many levels. Such links should be designed to make it easy for the student to move between subject content developed by the lecturer and the richness of the wider information environment as provided by the traditional library. For example, when a reading list is provided online, the student should not have to leave the course environment to conduct a search in the local library catalogue; rather, by clicking on the reading list item, the library holdings record should be displayed, giving the student the option to place a hold on the item. In the case of a journal citation, where licensing permits, the student should be able to click to the article without being prompted for further authorization.

Within such a learning environment a learner may choose to use a library portal for interaction with the broad, rich resource base the library offers plus the full spectrum of generic library services. The aggregated view of a library portal will complement the more selective subject-based portals. The generic library portal also should be presented from within other contexts across the institution's website where the library stands to benefit from the 'surprise' factor. For example, a link to a topical reading of the day, selected from local popular literature, might appear on the online cafeteria menu under the heading, 'Reward your mind as well as your stomach'. At the whole-of-library level there are many ways that librarians might transform their information literacy activities for online learners, but if these activities are to be credible, it is important that they are closely tied to the teaching and learning strategies of the institution. It is becoming possible to provide tools to allow learners to build up their own portfolio, perhaps against a predefined structure, to ensure that they progress through successive levels of competency in areas of information skills. While it is likely that students who start university in the next few years will have higher levels of information technology literacy, there are indicators that they will have reduced levels of general literacy and numeracy skills (Smith, 2000), increasing the need for library services that address the development of generic skills. This calls for closer partnerships with faculty to ensure that library services ultimately support learning outcomes.

A third approach is to develop a template that can be adapted for customization and integration within the online curriculum of a flexibly delivered unit. This is the aim of Discovery, a tool developed at the University of Ballarat in Australia (www.ballarat.edu.au/ubonline/showcase2001/discovery.html). Discovery was designed to support resource- and problem-based learning, critical enquiry and independent or collaborative research activities within courses delivered online. It facilitates access to key library services and resources, incorporating tools that support the desired use of information, for example, guidelines for constructing a bibliography using the relevant citation guide (Counsel, 2001)

Supporting teachers

At the University of Leeds, library staff undertook an assessment of the effectiveness of their integration strategies (Emly and Ryan, 2001). The Leeds Hybrid Library Project looked at the impact and value of integrating library and information resources into specific modules of the web-based learning environment. For a selected group of modules faculty librarians worked with the relevant academic to identify core resources. These generally included reading lists, customized searches, links to relevant databases and e-journals and specific search tools. Students had reservations about the hybrid library services. It was the academic staff who perceived the greatest value in the integrated approach, and this was largely because, in working with faculty librarians, they had become more aware of the potential of collaboration.

The teaching workforce in most universities is largely untrained for the instructional role. Lecturers and tutors need help, whether with reformatting their materials for the web or with redesigning the learning experience to incorporate student interaction with information. There is a range of services in most universities which might provide support. For libraries, an integrated approach means working with other learning support services and staff development teams towards common goals. Library-based instructional support services were the subject of an Association of Research Libraries survey of 121 member libraries in 2001 (Snyder, 2001). The survey found that, though many libraries were offering services to help faculty meet their teaching objectives, in most cases those services had been absorbed within the duties of existing staff and within existing budgets. As McLean (2000) notes, most universities are organized around a model where an individual teacher creates and delivers an individual unit of study. Cost structures are tied to the traditional model.

The INSPIRAL study demonstrated the widespread view that the current organizational structures of universities impede progress towards a more collaborative approach to creating online learning experiences that take advantage of a rich information environment.

Supporting researchers

Libraries have a history of working collaboratively with each other, particularly within industry sectors. A consortium of Australian university libraries is developing a research portal geared to the needs of staff and students of Australian universities. The Australian Academic Research Library Network (AARLIN) involves a range of research organizations, including the National Library of Australia, CAVAL Ltd (the State of Victoria's regional library co-operative network) and 19 universities. The project is funded through the Australian Research Council to 2003. The project's long-term vision is to develop a national virtual research library system that will provide access to the collections and resources of Australian academic libraries. AARLIN will provide portal services that may be delivered via the local web pages of any of its member organizations. AARLIN is just one of many Australian initiatives seeking to develop a national resource-sharing infrastructure; other examples may be accessed on the National Library's website (www.nla.gov.au).

In the UK the Resource Discovery Network (RDN) (www.vts.rdn.ac.uk/) is developing a series of subject portals that provide access to both high-quality materials freely available on the internet and the resources of commercial publishers. The RDN draws on the skills of a distributed network of subject specialists representing over 60 of the leading education and research institutions in the UK and provides a model for collective development of discipline-based resource portals. The RDN is developing technologies to allow content to be selectively served as part of a university's own web services, alongside other tools that contribute to building social capital across the campus community.

The idea behind the development of subject portals, yet to be realized, is that they will encourage scholars to interact with library services, commercial abstracts and indexes, gateways and other resource-finding tools relevant to their subject area. University students should be able to conduct searches either at the portal site or at a portal search box on their own institutional sites that would draw on a number of these services. The portal will use data about the user and the content to determine the most appropriate sources, and these can be presented according to the user's pre-defined preferences. A subject portal is a powerful community-building tool, allowing relevant information to be found quickly, shared and reused and added back into the pool. At the level of an individual university it improves the teaching–research nexus by making available the knowledge being created within the institution for class use. This ultimately has value in terms of promoting the university and enhancing its research profile.

As it becomes easier for faculty to create and share their own content, the management of intellectual property rights has become more problematic. Copyright in the digital environment raises some new challenges for libraries, not least because it is not a given that the processes to ensure proper rights management across the institution will be a library function. Harnad (2001) has been a leader in encouraging the academy to rethink scholarly communication for an internet-enabled world. Many libraries have set up digital repositories of research papers. There is no reason, as methods for distributed search and discovery improve, for each university to have its own e-print repositories.

A new learning and information environment

What are the components of an effective information-rich learning environment? What activities do librarians need to engage in to achieve progress towards putting these in place?

While there is a strong emphasis in the library literature on the need to work in partnership with academic colleagues, there is an increasing uncertainty over what the role of the librarian is in this alliance. How students are encouraged to interact with and use electronic resources is ultimately the responsibility of the faculty, not the librarian. It is the lecturer's responsibility to direct the information-seeking activities of students. The Conversational Framework espoused by Laurillard describes a more progressive model of teaching than the transmission model that libraries are used to serving. In this framework, teaching is 'an iterative dialogue between teacher and student(s)' (Laurillard, 2002). Through this dialogue learners are encouraged to understand the teacher's concept. Depending on the learning objects and the aspirations of the course, students can be helped to interact with resources. Through effective design and evaluation, the student is able to investigate relationships between texts or other digitized resources, guided by advice on what to look for and how much of a resource to use.

Twigg (2001) emphasizes the need to reconceptualize learning environments for a networked world. The library needs to rethink its contribution to supporting students. Returning to the question of the balance between being reactive and pro-active, for libraries the ultimate issue may be how to strike the balance between providing services that support the existing academic enterprise, which was built around a transmissional model of education, and providing leadership through transforming their services to fit the emerging model described by Laurillard. A learning management system can provide a frame for learning that is like a classroom. When librarians participate in the design of course materials, it is like adding a study hall to the classroom. And yet, a good campus library is much more than a study hall – it is an integral part of the formal education process at both the conceptual level and at the active, practical level. It meets the informal research and socialization needs of students, and it encourages students to extend their understanding by making new connections. It enables students to become more self-directed in their learning by providing diverse orientation experiences and ongoing opportunities for guidance, whether through bibliographic instruction, by peers or at the reference desk. The library makes a major contribution to the students' ability to participate in a campus community. The library culture motivates enquiry, discovery and investigation and is discipline neutral. Libraries provide a social context, without which information remains nothing but flat factual knowledge (Brown and Duguid, 2000).

Through using their library collection, students gain a sense of a body of knowledge and the divisions between disciplines and genres. It will be some time before digital libraries are able to supply learners with the range of online information that fairly represents the span of cultural and scholarly knowledge stored in university libraries. Educational theory suggests that learning requires cultural and social experiences and practice that can only happen through progressively engaging with a community. Online learning environments can certainly foster these experiences, but it still seems that a university should require that students are presented with a wider view of the world of knowledge than that typically presented within an individual course. As universities move towards teaching methods that better support students in 'the generic skills of scholarship' (Laurillard, 2002), where are librarians' efforts best placed?

These issues require further research. Arms (2000) says 'a digital library is only as good as its interface'. Different types of information and different formats require different interface designs, just as do different categories of users. A different look and feel for the interface will be required, depending on desired learning outcomes or the role of the user. Where the interface is provided by a proprietary learning management system, library services may be delivered via channels. Only by undertaking trials of a variety of delivery methods will it be possible to gain an idea of which work best for which type of situation. In the distance education literature there is some understanding of the relative effectiveness of different approaches, and it is possible to build on this. To foster the development of a general culture of learning across an organization, a library may need to use a variety of local interfaces, plus a generic interface which carries the institutional look and feel, while clearly branding the library service. There is an inherent tension between the information 'landscape' and 'brand-scape', and libraries face the dilemma of how to promote their role as service provider without confusing the users, who simply want some information that matches their needs.

Learners in a networked environment are able to fuse elements of information from different domains. The network reduces the time/place distance between learners and the appropriate information. Moreover, electronic information resources are increasingly available in multimedia formats and as hypertext-based products, which present learners with the capability to interact with content in entirely new ways and at new levels of granularity. Textbooks may be supplemented by 'learning-ware' incorporating web-based activities, simulations, case studies, visual images of concepts, even virtual laboratory experiments and virtual field trips. Access technologies are opening up new pathways, and to some extent this will in itself generate new uses of content. For example, many academic staff are presently unaware of the existence of repositories of Reusable Learning Objects. Most staff developers are used to authoring their own course materials and have not considered constructing course materials from learning objects created by others.

Part of the idea of an integrated learning environment is that the traditional distributed systems providing computer and information services across the university are utilized to support the individual user. To allow users to access and manipulate information to suit their needs requires a wider web-based information architecture. The vision of a managed learning environment places the user at the centre of a set of frameworks that includes the web, authentication, authorization, personalization, services and content, while supported by an underlying infrastructure. Lippincott (2002) uses the term 'cyberinfrastructure' to describe the emerging interrelationships between computing, networking, digital libraries and learners. The library sector is but one partner involved in the building of the new information infrastructure. Lippincott believes that librarians are well placed to contribute to the cyberinfrastructure. In particular they bring an understanding of mechanisms that are responsive to the changing needs of user communities, an ability to identify appropriate information sources, and the standards and methods that underpin searching for relevant content.

Conclusion

Through research and innovation librarians are seeking ways to enrich the online learning experience and exploit the inherent strengths of the internet. There is a myriad of appropriate ways to improve library services by embedding them into the online learning experience. The INSPIRAL final report (Currier, 2001) notes that there has been little evaluation of library effectiveness in delivering integrated services. Similarly, there is not a clear picture of how the role of the librarian is changing because of the shift to online delivery that is occurring across the higher education sector. Where libraries are experimenting with methods to support online learners, there are many new issues to consider – course design and quality, student perceptions of their learning experience, organizational change in higher education, interface design and standards for interoperability. Sharing experiences and expertise is critical to gaining a better understanding of how digital library technologies contribute to learning and teaching.

As in Greenstein's tentative view of the future 'digital library service environment' (Greenstein 2002), the place of the librarian is likely to continue be defined in familiar terms, as mediating between the distributed universe of information resources on one hand and, on the other, the needs of the user. A new feature is the emergence of internet-enabled communities of learners. More knowledge is required about the needs and expectations of the new generation of library users, both in regard to their use of internet-based information and their use of internet-based capability in being able to interact with those resources in completely new ways.

References

AARLIN Project
www.latrobe.edu.au/arlin/
(accessed 10 February 2002)

Appleton, M. (1999) Flexible delivery and the role of the librarian. Paper presented at QULOC, available at www.library.cqu.edu.au/publish/quloc-seminar.htm (accessed 2 April 2002)

Arms, W. (2000) *Digital libraries,* Cambridge, MA: MIT Press.

Barone, C. (2001) Conditions for transformation: infrastructure is not the issue, *EDUCAUSE Review*, (May/June), 40–7, available at www.educause.edu/ir/library/pdf/erm0133.pdf (accessed 12 April 2002)

Barth, C. D. and Cottrell, J. R. (2002) A constituency-based support model for delivering information services, *College & Research Libraries*, **63** (1), 47–52.

Beagle, D. (2000) Web-based learning environments: do libraries matter?, *College & Research Libraries*, **61** (4), 367–79.

Biggs, J. (1999) *Teaching for quality learning at university*, Buckingham: Open University Press.

Blackall, C. (2002) Rethinking information literacy in higher education: the case for informatics. Paper presented at VALA 2002, Melbourne, 6–8 February 2002, organised by the Victorian Association for Library Automation, Inc., available at www.vala.org.au/vala2002/2002pdf/10Blakal.pdf (accessed 12 April 2002)

Brophy, P. (2001) *The library in the twenty-first century*, London: Library Association Publishing.

Brophy, P. and MacDougall, A. (2000) Lifelong learning and libraries, *New Review of Libraries and Lifelong Learning*, **1**, 3–17.

Brophy, P., Craven, J. and Fisher, S. (1998) *The development of UK academic library services in the context of lifelong learning. Final report*, Manchester: Manchester Metropolitan University,

Department of Information and Communications, Centre For
Research in Library and Information Management, available at
www.ukoln.ac.uk/services/elib/papers/tavistock/ukals/ukals.
html#Heading73
(accessed 8 March 2002)

Brown, J. S. (2000) Growing up digital: how the web changes work,
education and the ways people learn, *Change*, (March/April),
available at
www.aahe.org/change/digital.pdf
(accessed 8 March 2002)

Brown, J. S. and Duguid, P. (2000) *The social life of information*,
Boston, MA: Harvard Business School Press.

Carlson, S. (2001) The deserted library: as students work online,
reading rooms empty out — leading some campuses to add
Starbucks, *The Chronicle of Higher Education*, (16 November), available at
http://chronicle.com/free/v48/i12/12a03501.htm
(accessed 12 April 2002)

Clark, J. (2002) Subject portals: a new information delivery model
to enhance teaching and learning. Paper presented at the
International Lifelong Learning Conference, 16–19 June 2002,
Rockhampton, Queensland, organised by Central Queensland
University, Rockhampton. Paper unpublished.

Cohen, D. (2001) Where are the libraries?, *CLIR Issues*, **23**, available
at
www.clir.org/pubs/issues/issues23.htm
(accessed 12 April 2002)

Collis, B. and Moonen, J. (2001) *Flexible learning in a digital world:
experiences and expectations*, London: Kogan Page.

Counsel, R. (2001) 'Discovery' Online: enhancing VET research and
information access. Workshop delivered at Net*Working 2001:
from virtual to reality, Brisbane, Queensland; an online learning
conference held 15–17 October 2001, organised by Flexible
Learning Advisory Group (FLAG), Australian National Training

Authority, available at
www.flexiblelearning.net.au/resources/1resources-s233.html
(accessed 8 April 2002)

Cowley, B. (2001) Tacit knowledge, tacit ignorance, and the future
of academic librarianship, *College & Research Libraries*, **62** (6),
565–83.

Currier, S. (2001) *INSPIRAL: final report*, available at
http://inspiral.cdlr.strath.ac.uk/
(accessed 12 April 2002)

Dempsey, L. (2000) Scientific, industrial and cultural heritage: a
shared approach, *Ariadne*, **22**, available at
www.ariadne.ac.uk/issue22/dempsey/intro.html
(accessed 6 March 2002)

Emly, M. and Ryan, C. (2001) Adding value to student learning:
integrating the hybrid library into the virtual learning environment, *New Review of Information Networking*, **7**, 225–35.

Great Britain. Department for Education and Employment. Press
Office (1999) Blunkett to transform post-16 learning. Press
release, available at
www.uuy.org.uk/projects/tecs98/wpaper/pr0630.htm
(accessed 26 April 2002)

Greenstein, D. (2002) Next generation digital libraries. Paper presented at VALA, Melbourne. Conference held 6–8 February 2002,
organised by Victorian Association for Library Automation Inc.
(VALA), available at
www.vala.org.au/vala2002/2002pdf/01Grnstn.pdf
(accessed 12 April 2002)

Harnad, S. (2001) The self-archiving initiative: freeing the referred
literature online, *Nature*, **410**, (26 April), 1024, available at
www.cogsci.soton.ac.uk/~harnad/Tp/nature4.htm
(accessed 12 April 2002)

Hawkins, B. L. (2000) Technology, higher education and a very
foggy crystal ball, *EDUCAUSE Review*, (November/December),
64–73.

Hernon, P., Powell, R. R. and Young, A. R. (2002) University library directors in the Association of Research Libraries: the next generation, *College & Research Libraries*, **63** (1), 73–90.

Joint Information Systems Committee. Distributed National Electronic Resource (2002) *DNER information environment: development strategy 2001-2005 (draft)*, available at www.jisc.ac.uk/dner/development/IEstrategy.html (accessed 6 March 2002)

Laurillard, D. (2002) Rethinking teaching for the knowledge age, *EDUCAUSE Review*, (January/February), 17–25.

Lippincott, J. K. (2002) Cyberinfrastructure: opportunities for connections and collaboration. Paper presented at VALA, Melbourne. Conference held 6–8 February 2002, organised by Victorian Association for Library Automation Inc. (VALA), available at www.vala.org.au/vala2002/2002pdf/31Lipnct.pdf (accessed 12 April 2002)

McLean, N. (2000) Library services for a managed learning environment. Paper presented at ALIA 2000. Capitalising on knowledge: the information profession in the 21st Century, Canberra, 24–26 October 2000, available at www.lib.mq.edu.au/conference/mclean/managed/index.html (accessed 12 April 2002)

McLean, N. (2001) Interoperability convergence of online learning and information environments, *New Review of Information Networking*, **7**, 27–42.

Queensland University of Technology (2002) *Course Materials Database*, available at www.lib.qut.edu.au/elibrary/courseworkcmd.html (accessed 14 May 2002)

Resource Discovery Network. (2002) *RDN Virtual Training Suite*, available at www.vts.rdn.ac.uk/ (accessed 1 May 2002)

Roberts, S. (2001) Transforming learning and teaching, transforming professional roles: Edge Hill and WebCT. Paper presented at Libraries Without Walls IV: the delivery of library services to distant users, Lesbos, Greece, 14-18 September 2001, organised by the Centre for Research in Library and Information Management (CERLIM), Manchester Metropolitan University.

Selby, M. and Young, C. (2002) *The Course Materials Database (CMD): your questions answered*, available at
www.sci.qut.edu.au/ag/acad_files/CMDyqa2.pdf
(accessed 14 May 2002)

Smith, K. (2000) New roles and responsibilities for the university library: advancing student learning through outcomes assessment. Paper prepared for the Association of Research Libraries, available at
www.arl.org/stats/newmeas/outcomes/heo.html
(accessed 12 April 2002)

Snyder, C. A. et al. (2001) *Instructional support services*, SPEC Kits, 265, Washington, DC: Association of Research Libraries, available at
www.arl.org/spec/speclist.html
(accessed 2 April 2002)

Twigg, C. A. (2001) *Innovations in online learning: moving beyond no significant difference*, New York: Pew Learning and Technology, available at
www.center.rpi.edu/PewSym1.html
(accessed 12 April 2002)

University of Bristol. Learning Technology Support Service (2002) *Why the anorak?*, available at
www.ltss.bris.ac.uk/anorak.htm
(accessed 8 March 2002)

11

It's just a click away, or is it? The challenge of implementing a global information network

Diane Kresh

Introduction

As a metaphor to frame a discussion of access to web and internet resources around the world, the story of Tantalus in Greek mythology is appropriate. Tantalus, king of Sipylos, was the son of Zeus and father of Pelops and Niobe. Although he was admitted to the society of the gods, his transgressions aroused their anger, and Zeus condemned him to suffer eternally in Tartarus. The legends disagree as to the exact nature of his infraction: one legend says that he had divulged divine secrets and stolen the gods' sacred food; another, that he had murdered his son Pelops and served his body to the gods to test their omniscience. As punishment, he was condemned to hang from the bough of a fruit tree over a pool of water. When he bent to drink, the water would recede; when he reached for a fruit, the wind would blow it from his reach. A further account of his punishment tells of a great stone hanging over his head, threatening to fall. One can see how the word 'tantalize' originated from his name.

While many of us in the business of providing information speak glibly about the world wide web (WWW) being everywhere, for more than 80% of the world, the web is nowhere, its accessibility a remote concept that falls to the bottom of the hierarchy of needs of daily life (World Economic Forum, 2001/2002). For many in the world the internet is like the fruit and water were to Tantalus – unreachable because of economic policies that keep people impoverished and une-ducated, government systems that oppress freedom of information, state monopolies on telecommunications systems that keep connec-tivity expensive, absence of electricity in rural areas.

Digital communication in this new information society is creating new resources in unprecedented numbers and making available the equivalent of billions of pages of words and pictures. Though libraries have continued to serve as entry points to the information superhighway, with so much information online and freely available we still ask ourselves who needs libraries or librarians. Some of the professional literature being published in the Western, industrialized world still abounds with articles that question the relevance of libraries and librarians when 'everything' appears to be available online. While this chapter will not settle the debate, perhaps it can place some perspective on the role librarians and library services can play in a world where the promise of global access to information, although not yet fully realized, is integral to the health and welfare of its inhabitants.

Who uses the internet?

In the USA the public turns to the internet for a variety of informa-tion needs: recreation, shopping, health and work-related research, online news and financial information. E-mail remains the most often used application, with 79% of internet users; and online shopping and bill paying are seeing the fastest growth. Low income users are the most likely to report using the internet at their local library to look for jobs. A recent study verified what many of us already assumed to be the case, that the internet has supplanted the library as a research tool for many online teenagers (The Pew Internet & American Life Project).

The Markle Foundation (2001) found that 63% of all Americans, and 83% of those who go online, have a positive view of the internet. The research further confirmed that the public identifies the internet primarily as a source of information – with 45% saying their dominant image of the internet is that of a 'library' as opposed to 17% who compare it to a 'shopping mall' or 'banking and investment office'. Despite the internet's popularity, a significant portion of those surveyed feel that the authenticity of the internet is of concern and that much of what is read on the internet must be questioned and indeed the proliferation of the 'instant expert' phenomenon has been well documented. According to the *New York Times* (Guernsey, 2000), 'an expert, it seems, is now an ordinary person sitting at home, beaming advice over the Internet to anyone who wants help'. The story of the 15-year-old teenager who filled out a form and became an official legal expert for an internet service represents the chilling downside of such entrepreneurial spirit. With 'experience' gleaned from watching years of court television, the young man found himself ranked number 10 out of approximately 150 experts (many of them actual lawyers) in AskMe.com's criminal law division (Lewis, 2001).

Although more users are online, study after study confirms that the internet is complex and hard to use. The best search engines cover only a third of the web; the rest is 'invisible' – hidden in databases that search engine spiders cannot penetrate (Price and Sherman, 2001). Current research has shown that enormous expanses of the internet are unreachable with standard web search engines. Only 16% of internet-based information can be located using a general search engine. The other 84% is information stored in databases. Unlike pages on the visible web, information in databases is generally inaccessible to the software spiders and crawlers that compile search engine indexes.

Differences between the web and libraries

The WWW currently hosts more than 8,745,000, sites (OCLC, 2002). The web also offers unprecedented opportunities for the creation and distribution of information in an unmediated and unrestricted environment. The digital environment, however, lacks the librarians, publishing houses and traditions of peer review that regulate the print environment. Without such collaborative regulation, information seekers are forced to wade through mountains of disorganized data in the hope of finding reliable, authoritative, historically significant and useful content. Edward Kerr (MacDonald, 2001) summarizes what many of us have personally experienced, that search engines 'may produce thousands of results, but often these are full of links to useless information, advertising banners and promotional garbage, and not to the information the person is looking for'. Yahoo!, while popular with library patrons, is seriously flawed as a reference tool (Cohen, 2001). A large percentage of its listings are commercial websites and not of high research value. Convenience rules the realm of the commercial Ask-A services (Parsons, 2001). They perpetuate the illusion that all answers to questions may be found on the internet by searching only what is available on the internet, not the vast collections found in libraries nor the thousands of library online catalogues that describe, organize and manage those collections.

In general, productive use of the internet is undermined by several factors, including:

- the absence of traditional ways of organizing information
- the collapse of distinctions between credible (or professional) and non-credible (or informal) knowledge producers
- the lack of high-quality, educational, diverse and socially important content online
- the web's ephemeral nature – internet information is often here today and gone tomorrow.

Libraries, with their vast collections of artefact-based knowledge, offer inestimable opportunities for information mining and employ reference and subject specialists whose knowledge is based on years of academic study and personal experience. Libraries are different from the internet in a number of other ways. For example, libraries:

- organize information using controlled vocabularies and other standards tools to make materials accessible
- evaluate materials carefully before selecting them according to documented policy statements and guidelines
- serve as society's memory by collecting and preserving print, non-print and digital material
- permit patron access through multiple communications options
- evaluate the needs of the patron through the reference interview
- promote inclusiveness and diversity and preserve cultural identities
- empower individuals by providing open access to resources and facilitating information literacy skills
- encourage a sense of community by providing a safe physical place for individuals to learn, discover, invent and renew.

In short, libraries add value. The hallmarks of libraries – structure and organization, in-depth subject expertise, sensitivity to the needs of patrons, community-vetted standards and best practices, commitment to free and open access to information and analogue collections – contrast strongly with the universe of unstructured, unverified and unmediated information on the internet.

The digital divide

There is another side to the story. The internet has affected the field of librarianship – if nothing else it has encouraged us to think of providing services in new ways. How much of an impact it has made cannot yet be determined, however, because internet 'penetration' is far from equal around the globe. Use of the internet also varies from country to country and from region to region. In Europe, for example, statistics show a range of usage from very high use in Finland (attributable to the extensive phone lines that became digital 15 years ago), to almost no use in Spain (*Europe*, 2001/2002). Indeed, use of the internet is so strong in Finland that it is regarded as an integral part of Finland's long-term strategy, known as the Finland 2015 Plan, the goal of which is to improve the knowledge, skills, resources, and networks of top-level Finnish decision makers in matters concerning the future of Finland.

At the other end of the spectrum in Europe the Spanish government published a report lamenting the poor level of internet use among its citizens and the low numbers of personal computers in Spanish homes. Statistics show that only around 16 % of Spaniards have access to the internet, putting Spain second from the bottom in European Union rankings. Another factor contributing to low use is the outdated infrastructure that exists in some areas of Spain. Analysts say that hundreds of thousands of people living in rural regions do not have access to telephone lines capable of handling internet signals.

Internet use in Central and Eastern Europe lags considerably behind the rest of Europe, and the barriers include everything from the poor state of the underlying communications infrastructure to the practice of per-minute charges for local calls and restrictive government regulations (Center for Democracy and Technology, 2001). The picture in Africa is even more bleak, as poverty remains the principal barrier to internet growth. As recently as 1999, and excluding South Africa, only one African in 9000 had access to the internet, while around the world the average is one in 40 (Editorial, 2000). In Africa e-mail is the most common digital form of communication, but the system in which it operates, while low in cost, results in delayed sending of messages, ranging from several hours to a few days.

Statistics published in the *Global Digital Divide Initiative annual report* (World Economic Forum, 2001/2002) offer a sobering picture of the wide gulf that exists between the information haves and have nots: industrialized countries with only 15% of the world's population are home to 88% of all internet users; less than half of the US households with annual incomes under $15,000 (representing 19% of the US population) will be online by 2005, etc. This report complements a series of reports issued by the US National Telecommunications and Information Administration, *Falling through the net: defining the digital divide* (2000), which confirm the gap between those with access to the internet and those without. As information technology plays an ever-increasing role in the lives of people around the world, a digital divide can impede the health of all communities, development of a skilled workforce, and ultimately the economic welfare of nations.

Although *Falling through the net* cites many positive gains in connectivity, the divide has not disappeared and in fact has expanded slightly in some cases, even while internet access and computer ownership are rising rapidly for almost all groups. For example, the data show that noticeable divides still exist between those with different levels of income and education, different racial and ethnic groups, old and young, single- and dual-parent families, and those with and without disabilities. Although the findings in this report show that there has been tremendous progress, there are still large sectors of American society that are not adequately connected. Each year being digitally connected becomes ever more critical to economic, educational and social advancement. Now that a large number of Americans regularly use the internet to conduct daily activities, people who lack access are at a growing disadvantage. Raising the level of digital inclusion by increasing the number of Americans using the technology tools of the digital age has become an important national priority.

However, the connectivity picture is changing. A report released in August 2001 by Dataquest, a research unit of Gartner Inc., has indicated that by 2003 Asia will pass the USA in number of internet users, making the Asia-Pacific region the world's largest internet market. According to the study, the Asia-Pacific region – including Japan, China, South Korea, India, Taiwan and other countries – will have

183.3 million internet subscribers in 2003. The USA will have 162.8 million internet subscribers, and Western Europe will have 162.2 million. The report further stated that the region will hit 248 million subscribers by 2005. Gartner Dataquest said three factors will drive rapid growth: lower access costs, improvements to infrastructure, and pent-up demand. The study went on to report that the country with the highest projected growth rate is India, which is expected to have 21.3 million internet subscribers by 2005. In 2005 China will be the largest internet market in the region, followed by Japan, South Korea and India (Gartner Dataquest, 2001).

Bridging the digital divide

While poverty remains the biggest obstacle for internet growth in Africa, there is hope offered by some of Africa's technology and government leaders. Ayisi Makatiani, founder of Africa Online, has observed that, through the internet and mobile phones, the educated are beginning to know more about the world and use that knowledge to influence what goes on around them. Ayisi attributed the outcomes of the elections in both Senegal and Ghana, in which presidents who had been in power for years were ousted, to the power of the mobile phone, the internet and the radio. 'You can't fake an election, any more', he said. 'When you count a ballot, ten people with cell phones are calling remote command centers. People go to cyber-cafes and send back tallies. So before the ballots come back to the capital, you know who has won' (Makatiani, 2001).

Two innovative US programmes begun in 2001 give rise to the hope that positive responses to the digital divide are possible and that such responses are not solely the responsibility of public libraries. The city of Houston, Texas (Swartz, 2001) has begun providing access to the internet and to software for word-processing and e-mail to all of its citizens through an arrangement with Houston-based Internet Access Technologies (IAT). IAT hopes to strike similar deals in up to 12 cities, including Chicago and Indianapolis, and may form partnerships with internet service providers such as AOL Time Warner and EarthLink.

The local branch of the public library has been taken to new heights as old meets new with the revitalization of e-library kiosks. E-library kiosks have sprung up in locales ranging from Baltimore, Maryland to San José, California, offering a range of information services, including homework help, e-mail access and reference assistance (Grenier, 2001).

Libraries as agents of change

From their inception libraries have provided local gateways to knowledge, have reflected the plurality and diversity of society and have supported the process of democratization. Information is power, and ensuring access to information is the role of the librarian, now more than ever. James Billington, in an address to the 2001 meeting of the International Federation of Library Associations and Institutions (IFLA), observed:

> The role of the librarian has become more, rather than less, important: to help learners of all ages make connections between print and electronic materials, and to help navigate through the sea of illiterate chatter, undependable infotainment and gratuitous sex and violence that is proliferating and that many say is the only real profit-making on the Internet. The Internet tends to feed upon itself rather than independently validate the material it transmits.
>
> (Billington, 2001)

The US-based Benton Foundation 1996 report, *Buildings books and bytes*, affirmed that library leaders want libraries of the future to be hybrid institutions that contain both digital and book collections and that the librarian will be the navigator who will guide library users to the most useful sources, unlocking the knowledge and information contained in the vast annals of the information superhighway. A similar report published by the UK Library and Information Commission (Great Britain. Library and Information Commission, 1997), *New Library: The People's Network*, summarized international perspectives on the development of the information society, finding points of view that were surprisingly similar regardless of the country of origin. Of the countries surveyed, all agreed that the information society would be one based on lifelong learning, that barriers to universal access to the information superhighway must be eliminated, and that policies must be developed globally and collaboratively in order to overcome these barriers.

Librarians intent on debating their purpose in the age of the internet would do well to review IFLA's *Core values statement* (2002), Article 19 of the UN's Universal Declaration of Human Rights (1948) and UNESCO's *Public library manifesto* (1994), each of which provides a strong philosophical basis to support the role of libraries in society. The *Core values statement* underscores the value of libraries in the digital age: people need free access to information and the provision and delivery of high-quality library and information services help guarantee that access. Article 19 states that 'everyone has the right to freedom of opinion and expression; this right includes freedom to hold opinions without interference and to seek, receive and impart information and ideas through any media and regardless of frontiers' (United Nations, 1948). The *Public library manifesto* proclaims UNESCO's belief in the public library as a living force for education, culture and information, and as an essential agent for the fostering of peace and spiritual welfare through the minds of men and women (UNESCO, 1994). Finally, the IFLA *Statement on libraries and intellectual freedom* stresses that a commitment to intellectual freedom is a core responsibility for the library and information profession and that libraries help to safeguard basic democratic values and universal civil rights (IFLA, 1999). It goes on to state that libraries should reflect the plurality and diversity in society and make materials, facilities and services equally accessible to all users without discrimination on the basis of race, creed, gender, age or any other reason.

Libraries respond to demands

Libraries are increasingly aware of the demands being made on them and are looking for ways to share the load. Market research conducted by the Library of Congress as part of its global collaborative digital reference service found that:

- resources of libraries are being stretched thin as demand increases and expectations of patrons continue to rise
- patrons are increasingly at a distance from the library and expect information to be delivered to them where they are
- reference requests are becoming more specialized, detailed and complex and require more effort to fulfil and access to resources beyond a library's walls
- there is strong interest in collaboration among libraries to share resources and extend better service to patrons
- interest in deploying e-reference services is very strong and will fundamentally change the profession of reference librarians (Library of Congress QuestionPoint).

In the rush to respond to the new expectations and needs many libraries around the world have taken the plunge and created digital reference services. Joe Janes and his colleagues at the University of Washington (Janes, 2000) found that 45% of academic libraries and 12.8% of public libraries now offer some type of digital reference service. These services, however, are often ad hoc and experimental. Highly publicized ventures like '24/7 Reference', Cleveland Public Library's 'Know It Now' and QuestionPoint created by the Library of Congress and partner libraries are evidence that such experiments are gaining traction as librarians revolutionize their services in order to stay relevant.

This is undeniably a watershed moment for libraries, an opportunity to create service visions that inspire and reflect the direction the profession must take to ensure that people continue to enjoy free access to information. This freedom is at the heart of their ability to participate fully in the democratic process. Libraries, along with schools, churches, civic associations, volunteer organizations, reading groups, and a host of other 'communities' have an important role to play in shaping and sustaining the social fabric (Preer, 2001). History and technology have combined to provide libraries with a unique opportunity to extend their reach, unfettered by time or place. History's greatest innovations have been about mass communication. The printing press linked communities and created a literary culture; the automobile linked urban and rural Americans and shrank the spaces between them; today, the web has spawned the potential, as yet unrealized, for a global community where, with a PC and a modem, a person can connect any time, anywhere with another person to share information, buy or sell a product, learn about each other – the opportunities are limitless.

How do librarians build on their age-old status as trusted advisers and create services that will both meet these new demands and revitalize the profession? How do we take the reference desk to cyberspace? With insightful leadership, the efforts of reference librarians can be pooled in a systematic way to fulfil an important mission of every library – to serve the needs of its community. In the global economy the community becomes the world, and it is as important for libraries around the world to join together to address the digital divide and ensure universal access to information as it is to initiate new programmes to exploit technology to its fullest.

It is worth remembering that libraries in this new information age are not just the 21st-century equivalents of the 19th-century passive storehouse. Their potential is limited only by the imagination of its creators. The proliferation of websites and the explosion of e-commerce and other e-services all argue for dynamic library programmes that will:

• employ technologies that provide better service and make library collections and resources more widely accessible and affordable to

patrons around the world and shrink the digital divide
- evolve into hybrid libraries (Rusbridge, 1998) and collect and create significant publications in both print and electronic formats so library and research collections continue to be universal and comprehensive
- build collaborations with both national and international institutions, including publishers and technology producers, to create shared assets enabling libraries to store, preserve, provide access to and expand their resources and ensure universal access
- develop and implement standards and best practices to facilitate interlibrary co-operation in the areas of information exchange between computer systems and to ensure that networked services are meeting user expectations
- create a culture of technical and strategic innovation so libraries can fulfil both traditional and new initiatives – a digital library's potential is limited only by the imaginations of its creators
- create opportunities for professional development and training to increase sensitivities to other cultures, build understanding and achieve true globalization (McSwiney, 2000).

Such library programmes are not, however, self-sustaining. They need people to manage them, and to do that effectively the human resources must be renewed through continuing education, including attendance at professional meetings and conferences. Library and information management programmes need to re-examine curricula to ensure that the tools they are giving graduates prepare them to teach information literacy, preserve and recover library materials in the event of a cataclysmic event, balance the need to know with the need to ensure privacy, embrace and support multiculturalism, and understand the ethics of information (Cox, 2001).

QuestionPoint

Launched by the Library of Congress in the spring of 2000, and with a growing membership of more than 250 libraries around the world, QuestionPoint, formerly the Collaborative Digital Reference Service (CDRS), enables libraries to help each other serve all of their users, no matter where the users are. Through this network of qualified professionals skilled in the art of locating, organizing and authenticating information, QuestionPoint combines the power and uniqueness of local collections with the diversity and availability of libraries and librarians everywhere, 24 hours a day, seven days a week.

The vision for QuestionPoint began with the recognition that, with all of the emphasis on creating digital collections, there had been little co-ordinated or collaborative focus in the USA on developing the public service potential of digital libraries. There was an increasing demand for public access to the internet in libraries and for remote access to collections and to language and subject expertise. In each case the reference librarian was faced with determining how best to satisfy the needs of both the on-site and the remote library user at a time when electronic resources were increasing at an astounding rate.

The collaborative nature was in evidence from the start as we collectively defined the business rules by which QuestionPoint would be developed and implemented. For example, it was agreed that:

- QuestionPoint is a membership organization
- the technology platform is built once to serve the membership as a whole and the 'cost' is shared among the members
- it is an open service and members need only internet access, a browser and e-mail to use it
- members are committed to high quality and are held accountable through service policies, membership certification and service level agreements to ensure that the brand lives up to market expectations
- QuestionPoint is an international service and allows for the broadest participation among types of libraries
- the business model is flexible and ensures that no one library or group of libraries bears all of the costs of establishing and sustaining it.

In summary, QuestionPoint:

- provides 24/7 access to global resources
- collects and catalogues knowledge for reference access
- adds value to information on the internet
- demonstrates flexibility in creating technical solutions
- creates partnerships with other service providers to deliver more effective information management and delivery tools
- balances the needs of member libraries to ensure the broadest participation.

As QuestionPoint expands globally and becomes a true 24/7 service, there are many issues that require examination. These include language and literacy in order to provide service to local populations in their own languages, navigating cultural and political sensitivities, e-commerce and trade agreements that may affect pricing models. As one example of the significance of these issues, by 2003 non-English language material will account for over half the content of the internet, so the service language is becoming a major issue. The solutions to these issues will determine the long-term success of QuestionPoint and similar services.

Conclusion: going where the patrons are

Libraries today are using technology to link those in need with credible and accurate resources. QuestionPoint is one of many experiments going on in the profession, innovative and creative experiments designed to make information available faster and more effectively to meet more specialized demands.

This is undeniably a time for librarians to reinvent themselves and to adapt their skills to the demands of the protean universe of information, a world that is paradoxically both so immediate and so unfathomable. Technological advances have created new opportunities for libraries, information managers, researchers and library patrons of all kinds. Indeed, the internet has created a fundamental change in the way people collect, manage and disseminate information and acquire knowledge. Instead of making a trip to the library, researchers turn first to the internet. Few would disagree, however, that many people continue to need the support of a trusted and accountable intermediary in accessing, interpreting and evaluating what is available online. Librarians perform the role of knowledge navigators partly because all people are entitled to equal access to information and knowledge. To remain current, librarians must adapt and not feel threatened by the pace of change but embrace it and use it to imagine new and more responsive programmes and services.

As the geographic borders that used to define us disappear into wireless networks of interoperability that mitigate the affect of time zones, the greater the likelihood that differences in cultural backgrounds and contexts will obscure the receipt and dissemination of information. The technical complexity of the global network will be far easier to overcome than will prevalent cultural and political prejudices and attitudes. It will take time, patience and well-placed global partnerships before we can shed our blinkers and interact responsibly with a world around us that is increasingly affecting each aspect of our daily lives.

No library can do it all alone. By interacting locally with libraries everywhere, the Library of Congress and its partner libraries bring control and context to the global and diverse world of information. Libraries around the world must initiate similarly aggressive and pro-active outreach programmes to remote users. We need to provide access to reference services in a way that is obvious and as convenient to the remote user as is access to the information itself. Libraries must form alliances to provide information services in the digital age. Anne Lipow (1999) reminds us that no single library can successfully provide reference service alone. It is rather a co-ordinated effort, with libraries doing more to connect indexed printed material with that which appears on the internet. Unlike library reference services tied to a print collection that is geographically fixed, electronic services are not fixed. Database providers may find a market for selling internet-accessible reference services; library consortia may consider the possibility of distributed virtual reference services where member libraries specialize in particular topics with a membership dispersion around the globe sufficient to provide a 24-hour service.

In closing, it is appropriate to refer to Sigrid Hannesdóttir's 'Global issues/local solutions: cooperation is the key' (2001). This paper articulates what libraries must do in order to sustain access to information. Hannesdóttir urges libraries to form networks with other stakeholders, content providers, policy makers and technology producers in order to share knowledge and information and there will be more division of labour, and more systematic approaches to the issues. Such stakeholders would include information policy makers, information technology developers, producers and suppliers of electronic information, staff of libraries and information centres, and the users.

Information policy makers. Libraries are dependent on the political atmosphere in each country; in particular they are dependent on national and international policy for libraries and for the citizens' rights to access to information and knowledge.

Information technology developers. IT developers and librarians need to work together to produce good, workable solutions to address issues such as machine–human interfaces and to create information portals and access gateways, standards for information storage and transfer, development of one-stop shops with access to information professionals from outside the library.

Producers and suppliers of electronic information. Librarians and creators of digital content must work together to select material for digitization and setting of priorities, to create standards for digitization, to develop licensing and pricing structures, and to address copyright and intellectual property issues.

Staff of libraries and information centres. These are the ones with the requisite skills required to navigate through the information world effectively and efficiently, and these skills must continue to develop and change, keeping pace with developments in information provision and access.

Users. These are the most important of the libraries' stakeholders: the students, lifelong learners and specialists who use library facilities to inform, discover, and create.

Dr Hannesdóttir envisages that each library, independent of type or local clientele, will be part of a network, first and foremost a local network, but also a regional and eventually a global network where each institution assumes a certain responsibility in provision of information. Libraries in such a network will provide a one-stop shop open 24 hours, seven days a week, where all information is available. The network will support quality-controlled portals where the user can access information that has been checked for accuracy and quality, but also open access where the user can search for any kind of information publicly available and create customized mailboxes where individuals can register their interest profiles and receive current awareness information on new sources from the internet.

Behind Dr Hannesdóttir's vision of the library of the future, behind QuestionPoint and other networked services, are highly trained librarians and subject specialists who make it possible to reach the office or working space of each user, independent of where he or she is located. The explosion of online information, the popularity of the internet and commercial search engines, and the increased interdependency of access to information and economic health and prosperity have all required that librarians look afresh at their profession. New demands and expectations have emerged from the overwhelming amount of information now available. The requirement to deliver information to the remote user at point of need has encouraged the creation of many innovative programmes linking new technologies with traditional library services. Librarians now have an historic opportunity to adopt a new service paradigm and, in so doing, to demonstrate their relevance in the modern world. More important, they can use the new technology to provide a fundamental mission-based service that exceeds what might only have been a dream ten years ago, before the miracle of the world wide web. Technology provides the tools, but the librarian has the road map. In the global information society one does not work without the other.

References

Benton Foundation (1996) *Buildings books and bytes*, Washington, DC: Benton Foundation, available at www.benton.org/Library/Kellogg/buildings.html (accessed 10 April 2002.

Billington, J. H. (2001) Humanizing the information revolution. In 67th International Federation of Library Associations and Institutions Conference, Boston, MA, August 21, *IFLA Journal*, 2001, **27** (5–6), 301–7, available at www.ifla.org.sg/IV/ifla67/pages/183-129e.pdf

Center for Democracy and Technology (2001) *Bridging the digital divide*, Washington, DC: Center for Democracy and Technology, available at www.cdt.org/international/ceeaccess/ (accessed 10 April 2002)

Cohen, L. B. (2001) Yahoo and the abdication of judgement, *American libraries*, **32** (January), 60–2.

Cox, R. (2001) The day the world changed: implications for archival library and information science education, *First Monday*, **6** (12) (3 December), available at www.firstmonday.dk/issues/issue6_12/ (accessed 10 April 2002)

Editorial (2000) *Daily Nation*, (24 June). Information mentioned in Rao, R. (2001) Digital divide wider than rich–poor gap, *The Tribune*, 21 May, available at www.tribuneindia.com/2001/20010521/login/main6.htm

Europe: magazine of the European Union (2001/2002) **412** (December/January), Special report on Internet use in the EU's fifteen members, 36–46.

Finland 2015 Plan (2000) Finnish National Fund for Research and Development (SITRA), available at http://194.100.30.11/suomi2015/ (accessed 10 April 2002)

Gartner Dataquest (2001) Gartner Dataquest: APAC will overtake US in subscriber figures by 2003, available at www.internetnews.com/infra/article/.php/10693_861101 [7 August] (accessed 10 April 2002)

Great Britain. Library and Information Commission (1997) *New Library: the People's Network*, London: Library and Information Commission, available at www.ukoln.ac.uk/services/lic/newlibrary/full.html (accessed 10 April 2002)

Grenier, M. P. (2001) E- branch library machines help to bridge digital divide, *Wall Street Journal*, (12 July, also available at www.digitaldividenetwork.org/content/news/index.cfm?key=413

Guernsey, L. (2000) Suddenly everybody's an expert, *New York Times*, (3 February), Sec G, 1.

Hannesdóttir, S. K. (2001) Global issues/local solutions: cooperation is the key. In *11th Nordic Conference on Information and Documentation, Reykjavik, Iceland*, NORDINFO: Helsingfors, Finland, available at www.bokis.is/iod2001/papers/Hannesdottir_paper.doc

IFLA (1999) *Statement on libraries and intellectual freedom*, The Hague: IFLA, available at www.faife.dk/policy/iflastat/iflastat.htm (accessed 10 April 2002)

IFLA (2002) *Core values statement*, The Hague: IFLA, available at www.ifla.org/III/intro00.htm#3 (accessed 10 April 2002)

Janes, J. (2000) Current research in digital reference: findings and implications. Presentation at Facets of Digital Reference: The VRD 2000 Annual Digital Reference Conference, 17 October

2000, Seattle, WA, available at
www.vrd.org/conferences/VRD2001/proceedings/janes.shtml
(accessed 10 April 2002)

Lewis, M. (2001) Faking it: The internet revolution has nothing to do with the NASDAQ, *New York Times Magazine*, (15 July), Sec 6, 32.

Library of Congress QuestionPoint
www.loc.gov/rr/digiref/
(accessed 10 April 2002)

Lipow, A. (1999) Serving the remote user: reference service in the digital environment. In *Strategies for the next millennium: Proceedings of the Ninth Australasian Information Online & On Disc Conference and Exhibition, Sydney Convention and Exhibition Centre, Sydney, Australia 19–21 January 1999*, available at www.csu.edu.au/special/online99/proceedings99/200.htm
(accessed 10 April 2002)

MacDonald, T. (2001) Study: Internet rage hits the information highway, *NewsFactor Network*, (9 April), available at www.newsfactor.com/perl/story/8806.html

McSwiney, C. (2000) Think globally! In *Australian Library and Information Association (ALIA) 2000 Conference. Capitalising on knowledge: the information profession in the 21st century, 24–26 October 2000*, Canberra: Australian Library and Information Association, available at www.alia.org.au/conferences/alia2000/proceedings/
(accessed 10 April 2002)

Makatiani, A. (2001) Ayisi: founder, Africa Online, *Fortune Magazine*, **44** (11) (26 November), 108.

Markle Foundation (2001) *Toward a framework for internet accountability*, New York: Markle Foundation, available at www.markle.org/index.stm
(accessed 10 April 2002)

OCLC (2002) *The WEB Characterization Project web statistics*, Dublin, OH: OCLC, available at

http://wcp.oclc.org/
(accessed 10 April 2002)

Parsons, A. M. (2001) Digital reference: how libraries can compete with Aska services, *Digital Library Federation Newsletter*, **2** (1), (January), available at www.diglib.org/pubs/news02_01/RefBenchmark.htm (accessed 10 April 2002)

The Pew Internet and American Life Project
www.pewinternet.org/reports/reports.asp?Report=39
(accessed 10 April 2002)

Preer, J. (2001) Where are libraries in bowling alone?, *American Libraries*, **32** (September), 60–2.

Price, G. and Sherman, C. (2001) *The invisible web: uncovering information sources search engines can't see*, Medford, NJ: Information Today.

Rusbridge, C. (1998) Towards the hybrid library, *D-Lib Magazine*, (July/August), available at http://mirrored.ukoln.ac.uk/lis-journals/dlib/dlib/dlib/july98/rusbridge/07rusbridge.html (accessed 10 April 2002)

Swartz, J. (2001) Houston citizens get free e-mail, project attempts to close 'digital divide', *USA Today*, (20 August), 1A.

United Nations (1948) *Universal declaration of human rights*, New York, Article 19. Also available at www.un.org/Overview/rights.html

UNESCO (1994) *Public library manifesto*, Paris: UNESCO, available at www.ifla.org/documents/libraries/policies/unesco.htm (accessed 10 April 2002)

USA. National Telecommunications and Information Administration (2000) *Falling through the net: defining the digital divide*, Washington, DC: National Telecommunications and Information Administration, available at www.ntia.doc.gov/ntiahome/digitaldivide (accessed 10 April 2002)

World Economic Forum (2001/2002) *Global Digital Divide Initiative annual report*, Washington, DC: World Economic Forum, available at www.weforum.org/pdf/Initiatives/Digital_Divide_Report_2001_2002.pdf (accessed 10 April 2002)

Part 5

COLLECTION MANAGEMENT

12

Evaluating digital collections

Alastair G. Smith

Introduction

While true digital libraries are still the stuff of research and speculation about the possible future, most library collections contain, or
provide access to, some digital materials. The hybrid library has
become the norm (Oppenheim and Smithson, 1999), and librarians
must extend their evaluation techniques to digital collections.

What is included in the concept of digital collections? The traditional role of the library has been threefold: to share expensive
resources, to organize artefacts and concepts, and to bring together
people and ideas. In the digital environment these roles exist, but
change somewhat. Sharing of digital resources can become easier, since
bytes can be easily and exactly copied and made available instantly at
remote locations. Library collections are not limited to what is held in
a building, but can include materials that are accessed remotely.
However, this very ease of sharing raises intellectual property issues
which bring traditional library practices into question. Intellectual organization of a digital collection is not limited by the key characteristic of
physical objects (that they can only be placed in one subject area), but
at the same time intellectual control of these digital objects, whose content and location can change dynamically, is challenging.

The digital environment offers exciting new ways in which users of information can relate to each other and to authors, but it also threatens traditional concepts of the reviewing process and the authority of information. Digital collections may include materials that are accessible from the library, but not necessarily overtly selected. Materials that are accessible, but not necessarily selected, include: commentary by readers on digital works; materials that have similar content to others in the collection, but differ in format or view, and may have been generated dynamically in response to a specific user request. In the digital environment the same data may be accessed in multiple views, as 'multiple structures simultaneously' (Peters, 2000a), so there is not a fixed view of the content. All these characteristics of digital collections pose problems for evaluation.

Bound up with the concept of a digital collection is the concept of the digital library. Borgman (1999) points out that the term 'digital library' is used in at least two senses: in the computer science research community digital libraries are viewed as content collected on behalf of users, while in the library practitioner community digital libraries are seen as institutions providing a range of services in a digital environment. Most digital library projects fall into the first class, while speculation about future developments concentrates on versions of the second.

As digital collections and digital libraries emerge as the norm for various sectors of the information world, the evaluation of the collections of resources becomes an increasingly important issue. This chapter reviews some evaluation methods proposed in the literature and proposes a checklist of criteria that can be used in relation to digital collections. Further discussion is devoted to such related issues as evaluation of full-text services, databases and the web.

Evaluation of digital collections

The plentiful literature on evaluation of digital libraries and collections reflects a wide range of approaches. Broadly, evaluation efforts fall into four areas:

- content
- users and objectives
- services, performance measures
- user interface.

In all of these areas a central theme is the adaptation of traditional collection management practice to the digital environment. Nisonger (2001) highlights differences: in the digital environment access is more significant than possession, so collection management relates to selection of materials that may be made available virtually, rather than a collection of resources that will form part of a discrete physical collection. Nisonger points out that the availability of the internet influences the practice of evaluation in collection management in two ways: at the micro evaluation level it raises the challenge of evaluating internet resources for inclusion in the virtual collection; at the macro evaluation level it includes the evaluation of conventional library collections using internet resources, for example by using access to other library catalogues to compare collection strengths.

A key difference between digital and conventional collections is that digital collections must be considered as systems and not just collections of materials. The utility and user experience of digital collections lie not just in the content of the collection, but in how it can be used. Evaluation of a digital collection covers more than simply the content and includes performance (speed, reliability) of the computer systems, the information retrieval facilities, user interface, display formats, etc. According to Greenstein (2000), 'services are the distinguishing characteristic of a digital library'. Evaluation of digital collections requires a more multidisciplinary approach, and we need to draw on background from other disciplines, for example information systems, computer science and human–computer interface studies (Buttenfield, 1999; Jones, Gay and Rieger, 1999).

The view of a digital collection as a system has led to a number of evaluations of digital collections that have been in the nature of user studies. An example of a user evaluation is that by Hill et al. (2000).

They considered the Alexandria Digital Library of geospatial data from the point of view of various potential user groups of the collection, e.g. earth scientists, information specialists and educators. Evaluation methods included user surveys, ethnographic studies and a classroom study.

In the educational environment digital collections are being integrated with teaching resources, particularly in online and web-based teaching. This means that evaluation should consider the value of the materials as teaching and learning resources. Borgman et al. (2000) considered the use of the Alexandria Digital Earth Prototype (ADEPT) resources in conjunction with undergraduate education, and this evaluation specifically included learning outcomes. Marchionini (2000) looked at the use of the Perseus Digital Library in education. Sukovic (2001) considered the increasing importance in scholarship of electronic texts, which enable textual analysis methodologies that are not possible with traditional printed works.

Possibly the most important influence on user reactions to digital collections in the future will be the increasing expectation of 24/7 information service available digitally through the internet (that is, available 24 hours a day, seven days a week). Lipow (Lipow et al., 1998) points out that this increasing expectation is one of the drivers behind virtual reference services.

Despite the importance to the profession of developing evaluation methods for digital collections, some commentators rightly question whether digital collections are mature enough to evaluate (Peters, 2000a; Saracevic and Covi, 2000). In a rapidly developing area rigid standards or criteria could limit future developments (Saracevic and Covi, 2000) People today are using digital collections in ways unanticipated by the collection builders; for example, textual collections open up possibilities for large-scale textual analysis that would not be possible with conventional materials. Standards are still evolving – Jones, Gay and Rieger (1999) found varying metadata standards in the collections that they evaluated, because of the differing requirements of the collections, which failed to share enough metadata classifications. The field may be one appropriate for 'meta-assessment' (Peters, 2000b), where assessment is of the elements, basic conditions and needs of digital collections, rather than of specific collections. In a similar vein Buttenfield (1999) suggests assessing not just specific digital libraries, but evaluation methods themselves (the 'double loop paradigm').

Methods and criteria

I will review some evaluation methods proposed in the literature, and the following section will present a checklist of criteria that can be used in relation to digital collections.

The EQUINOX Project is a long-running European Union initiative to develop library performance and quality management measurements, and a software package that helps libraries to collect and analyse these (Brophy and Clarke, 2001). The 14 EQUINOX evaluation measures are:

1 percentage of the population reached by electronic library services
2 number of sessions on each electronic library service per member of the target population
3 number of remote sessions on electronic library services per member of the population to be served
4 number of documents and entries (records) viewed per session for each electronic library service
5 cost per session for each electronic library service
6 cost per document or entry (record) viewed for each electronic library service
7 percentage of information requests submitted electronically
8 library computer workstation use rate
9 number of library computer workstation hours available per member of the population to be served
10 rejected sessions as a percentage of total attempted sessions

11 percentage of total acquisitions expenditure spent on acquisition of electronic library services

12 number of attendances at formal electronic library service training lessons per member of the population to be served

13 library staff developing, managing and providing ELS and user training as a percentage of total library staff

14 user satisfaction with electronic library services.

The EQUINOX measures illustrate the problem of translating existing library evaluation measures into a digital environment. They largely relate to the model of a public or academic library providing access to digital services within its walls, and to remote users. It is not clear how these measures would relate to standalone digital collections, for example the Perseus Digital Library or the Making of America collections. To some extent the measures relate to a central computing model, where the library is responsible for total access and telecommunications. The EQUINOX measures relate very much to measurable quantities, avoiding qualitative measures such as the quality of content, and detailed aspects of user satisfaction. However EQUINOX does provide viable quantitative measures that could be used for comparison between libraries, and to track development of a library's digital services and place it on a continuum from traditional, to hybrid, to digital. In the USA McClure (1999) is pursuing parallel initiatives to evolve criteria for cross-library comparison.

Saracevic and Covi (2000) conclude that evaluation must consider two aspects: the construct aspect to define what is being evaluated, and the context aspect to define the level of evaluation, objectives, criteria, measures, methodology. They also identify a number of candidates for evaluation:

• digital collections, resources
  – selection, gathering, holdings, media
  – distribution, connections, links

- — organization, structure, storage
- — interpretation, representation, metadata
- preservation, persistence
- access
  - — intellectual
  - — physical
  - — distribution
  - — interfaces, interaction
  - — search, retrieval
- services
  - — availability
  - — range of available services, e.g. dissemination, delivery
  - — assistance, referral
- use, users, communities
- security, privacy, policies, legal aspects, licences
- management, operations, staff,
- costs, economics
- integration, co-operation (with other resources, libraries, or services).

This provides a wide range of aspects that can be considered for evaluation of a particular collection, although not all will apply to a particular collection or a specific evaluation project.

Jones, Gay and Rieger (1999) suggest three areas of evaluation: backstage concerns such as metadata and intellectual property; content issues such as collection maintenance and access; and usability. Kilker and Gay (1998) stress the implications of the social and cultural background of the library users and administrators. Kantor and Saracevic (1999) and Shim and Kantor (1999), drawing on the techniques of data envelopment analysis, suggest a measure of impact: the weighted logarithmic combination of the amount of time that users spend interacting with the service, combined with a Likert scale indication of the value of the service in relation to the time expended.

Specific elements for evaluation

Costs. Saracevic and Covi (2000) consider measures of effectiveness, efficiency, cost-effectiveness as being valuable indicators in an evaluation exercise.

Services. Usability is considered by Buttenfield (1999), emphasizing the importance of the interface in a digital library. To many users the interface is the digital library; on the other hand a traditional library can be used without the 'interface' of the reference librarian. Buttenfield advocates usability testing through interviews, observation, etc. However, there are problems with surveys: users of digital collections are not tolerant of delays introduced by participating in and administration of questionnaires, etc.

Interface. Hill et al. (2000) produced a list of requirements for evaluating the user interface: search functions, session management, result display, user workspace, holdings visualization, user help, usability, data distribution. Smith (2000) examined the search features of a number of typical digital library projects, and found that they were relatively unsophisticated: for instance few included controlled vocabulary, proximity searching, browsing of indexes, and refining of initial searches.

Access. Access is a significant factor in the effectiveness of digital libraries. Bishop (1998) found that authentication and registration procedures constituted a significant barrier that has tended to be unanticipated by digital library designers. Retrieval is another aspect of access. Lesk (1997) makes the point that despite advances in retrieval systems and their evaluation, barriers exist for those with low information technology skills, and that there is potential for the development of more sophisticated interfaces. A key issue is how 'item quality' can be represented in search systems. Existing search systems have no problem recovering large amounts of information, the problem users confront now is selecting quality information.

The advent of online reference services adds another aspect to access evaluation. These range from simple e-mail links, to reference librarians through HTML forms that structure an information query, to real-time reference interactions guided by purpose-designed software. Carter and Janes (2000) report on unobtrusive evaluation of the Internet Public Library. They point out that the text-based interaction of most virtual reference services creates evaluation opportunities that are not available for evaluating conventional services. In a similar vein, Janes (2000) describes a survey of academic, and subsequently public, library virtual reference services. This considers factors such as whether the service is linked from the library's web page, what intake methods are used to collect queries, what stated policies are present, whether there are technological barriers for users, and how FAQ pages are provided. Closely related to this approach is the need for the evaluation of question-answering services (such as 'Ask-A' expert referral services) to look at such criteria as accuracy, time to answer, service orientation, own evaluations of service, nature of responses and handling of out-of-scope questions.

Intellectual property issues. Most (historical) digital collections have avoided intellectual property issues by restricting the scope to public domain, out-of-copyright works (Jones, Gay and Rieger, 1999). Intellectual property issues do loom large, particularly as collections move from the realm of experimental digital projects to being inherent components of library service. Digital collections have significant opportunities for automatic rights management, however.

Quantity and quality. There is a critical mass of quantity appropriate to the use being made of the collection; 250,000 images may be insufficient if the scope is all images from all time periods and cultures; 500 images may be sufficient to give lay users an impression of the breadth of European art (Jones, Gay and Rieger, 1999). Optimal technical quality has been emphasized, but the importance of record quality is context dependent. For web use, low-resolution images of, say, 72 dpi may be optimal – text images need not be at high resolution for the content to be understandable.

An evaluation checklist for digital collections

The following list attempts to synthesize a range of criteria for evaluating digital collections, some of which have been noted above. Not all the criteria will be applicable to a given evaluation exercise or to a given digital collection.

1 Content
    a   Selection of materials
- Scope
- Coverage
- Influence of social and cultural background of selectors
- Size of collection
- Relation to library's overall selection policies

    b   Currency
    c   Accuracy
- Bias of selection, and of content

    d   Metadata
- Interpretative information
- Descriptive metadata
- Intellectual access metadata

    e   Formats
- PDF
- HTML
- XML
- TEI
- Images

    f   Preparation of materials
- Scanning resolution
- Accuracy of OCR

    g   Teaching and learning values
    h   Intellectual property
- Identification of copyright holders and other rights management information

2 Access

a Reference services
- Online availability
- Response time
- Hours of availability
- Quality of material provided (accuracy, authority, currency, etc.)
- Referral to experts
- Handling of out-of-scope questions

b Access over time: preservation
- Physical preservation
- Preservation of means of access
- Persistence of identity, address, locational information
- Preservation of view

c Barriers to access
- Logon procedures, security and access
- Bandwidth
- Service reliability
  - General internet access
  - Constraints imposed by library, e.g. number of telnet sessions

3 Services

a Users
- Nature of target user population:
  - Interests
  - Technological ability
  - Availability of network access to target users
- Percentage reached, level of use
- Workstation use within physical library
- Remote access:
  - From within organization
  - From members of organization working from home
  - From external users

- User satisfaction:
  - Content related
  - User interface
  - Value of service in relation to effort spent
- Privacy of user information

b   Training and support
- Number and nature of training sessions offered:
  - Content- and searching-related skills
  - IT-related skills
  - Creating resources
- Percentage of users attending training sessions
- Percentage of staff trained
- Percentage of staff devoted to training
- Documentation:
  - Software
  - Content
  - Reference
  - Context sensitive
- Ongoing support: newsletters, system announcements, notification of changes

c   Integration into other services
- Online teaching environments
- Referral to other resources
- Global searching of separated collections

4 Interface
a   Browsing
- Search indexes
- Results

b   Searching
- Fields indexed
- Operators available: Boolean, proximity, truncation, etc.
- Refining of initial search

c   Retrieval and display

- Formats, multiple levels of detail: citation, abstract, full text
- Display of search results: ranking by relevance, time, author, etc.
d   Session management
- Displaying and retrieving previous sessions and searches

5 Costs
a   Cost–benefit of services and resources
b   Acquisitions budget, percentage allocated to digital resources
- Permanent acquisitions of digital materials
- Licence fees for access to digital resources
- In-house development of digital resources
c   Staff resources devoted to the digital collection and to digital access
- Available skills, need for outsourcing
d   Costs per user
- Session
- Document
- Enquiry

Evaluation of full-text and database services

An important type of digital collection used in libraries is commercial full-text and bibliographic databases. Over the last decade bibliographic databases that provide analytical coverage of, for example, journals and conference proceedings, have extended their content to include the full text of the documents. Most libraries moving to the hybrid digital library model are using these services, often cancelling the print equivalents of publications available through the full-text database. Evaluation of these services is an important part of evaluating the library's overall collection, and various evaluation approaches have been tried.

Peter Jacso (1998) evaluated the coverage by Dialog databases of a range of library and information science journals, pointing out that limitations in the searching capabilities created problems in determining the level of coverage. Davidson, Salisbury and Bailey (2000), in evaluating the two full-text services PA Research 2 and WilsonSelect, compared numbers of journals in subject disciplines, effectiveness of searching and delivery options. Maquignaz and O'Neil (2000), in evaluating the Science Direct full-text service, found that cost and lack of access to back issues were problems. Still and Kassabian (1999) found that breadth and depth of coverage, manipulation of results and ease of searching were important in evaluating full-text database services.

A new form of full-text content in libraries is e-books (Badke, 2001; Gorman and Miller, 2001). E-books are generally leased from a supplier, and made available to users either over a network to display on the user's computer, or in some cases downloaded to a specialist reader or a PDA. Because of the complexities involved in evaluating materials available in different ways, models for the acquisition and availability of e-books are still evolving and are not yet available for application.

Evaluation of the web as a digital collection

For a long time bibliometric measures have been used in the evaluation of conventional materials and collections. Borgman et al. (2000) indicate that citation indexes have been influential in evaluating scholarly writing: illustrating how the social life of documents can be used to evaluate information for a variety of purposes. The potential for similar quantitative studies to be made of web-based materials is of significant research interest and has clear implications for the evaluation of digital collections. One name that has been applied to the extension of bibliometric methods to the web is 'webometrics' (Almind and Ingwersen, 1997).

An example of a webometric approach is the calculation of Web Impact Factors, which utilize the similarity of web hyperlinks to citations in print literature. Proposed by Ingwersen (1998), Web Impact

Factors are based on the well-established journal impact factors calculated from the ISI citation indexes. In simple terms, a Web Impact Factor is the ratio of links made to a site compared with the number of indexable pages at the site, and can be regarded as a measure of the extent to which the web community values the site by providing links to it. Web Impact Factors have given rise to a number of studies (Smith, 1999). Smith and Thelwall (2002) have examined the impact of Australasian university sites, using both a conventional search engine and a dedicated crawler to gather data. Various web-based measures were compared with research rankings of the universities. At this stage it appears that, while Web Impact Factors relate in some respects to other measures of the importance of sites, they are clearly different, and more research is required into the significance of linking behaviour.

In a sense the world wide web is another digital collection that a library can provide access to. However it is probably truer to regard the web as an access mechanism, through which libraries can select specific collections relevant to their mission. The criteria developed above can be used to apply to web-based digital collections, but there is also a strong literature relating to these resources in particular.

The importance of evaluating web content has been recognized for some time; there has been significant work on how people decide on web quality. Lists of criteria (Smith, 1997; Smith, 2001) include authority, accuracy, currency, workability, etc. Rieh (2002) suggests that users of the web make two types of judgement: information quality and cognitive authority (whether the information is from recognized sources). This suggests that digital libraries should facilitate the user's ability to identify the organization from which the information comes: to search for 'sources' as well as 'information objects'; that search engines should facilitate the display of clues to the information quality; and that web designers should include cues as to the cognitive authority of their pages.

Wathen and Burkell (2002) consider the issues of credibility in judging web-based information. They propose a model in which users first evaluate surface credibility (appearance, usability, organization), then evaluate message credibility (source expertise, trustworthiness, credentials and content, relevance, currency, accuracy). This appears to offer an innovative and relatively complete approach to the evaluation of web-based information and is an approach worth testing in some detail.

Evaluation of search engines and directories

A number of attempts have been made to evaluate web search engines; the parameters are rather different from those used to evaluate conventional database retrieval systems. Schwartz (1998) and Oppenheim et al. (2000) review some of the issues: the concepts of recall and precision need to be revised for relevancy-ranked searches which are the norm for web-based search engines; the internal algorithms used by the search engines are regarded as proprietary, etc. Monopoli and Nicholas (2000) evaluated the subject-based resource directory SOSIG using an online questionnaire to gain information on users' background and their experience of using the service. This seems to be a fruitful approach worth further exploration.

Conclusion

The evaluation of digital collections is a many-faceted problem. There is the problem of identifying the range of digital collections. These range from experimental digital library collections through full-text databases based on well-established indexing and abstracting services to the range of resources available through the world wide web.

A strong reason for the development of digital collections is the potential for remote access, but this creates problems in defining the user group and mission of the collection. A traditional physical collection is likely to have a more tightly defined and controlled user body. Similarly, a library is not restricted to its holdings in pursuing a digital collection policy, but can provide its users with access to resources that are available remotely. However, remote access lessens the control the library has over intellectual content, collection management and preservation.

More so than a conventional collection, a digital collection is more than content. Services, user interface and reliability of access are important aspects of the user experience of the collection, and indeed are in danger of overshadowing content in any evaluation. User expectations may well drive the development of digital collections and associated services that are available remotely.

In the academic area digital collections are assuming importance as teaching and learning resources, and also for research. This means that libraries must investigate ways to integrate digital collections with systems designed to support these objectives. The evaluation of digital collections must consider how easily the collection can be integrated with, for example, the organization's online teaching environment, or how texts can be transferred to analytical tools for linguistic research.

Because digital collections are still a relatively new feature in the information landscape, it will be some time before definitive methods of evaluation evolve. In the mean time there is scope for not just the development of evaluation methods, but also higher level initiatives to develop 'meta-assessment' so that we can recognize effective methods of evaluation when they appear.

References

Almind, T. C. and Ingwersen, P. (1997) Informetric analyses on the world wide web: methodological approaches to 'webometrics', *Journal of Documentation*, **53** (4), 404–26.

Badke, W. B. (2001) Questia.com: implications for the new McLibrary, *Internet Reference Services Quarterly*, **5** (3), 61–71.

Bishop, A. P. (1998) Measuring access, use, and success in digital libraries, *Journal of Electronic Publishing*, **4** (2), available at www.press.umich.edu/jep/04-02/bishop.html (accessed 6 April 2002)

Borgman, C. L. (1999) What are digital libraries? Competing visions, *Information Processing and Management*, **35** (3), 227–43.

Borgman, C. (2000) *From Gutenberg to the global information infrastructure: access to information in the networked world*, Cambridge, MA: MIT Press.

Borgman, C. L. et al. (2000) Evaluating digital libraries for teaching and learning in undergraduate education: a case study of the Alexandria Digital Earth Prototype (ADEPT), *Library Trends*, **49** (2), 228–50.

Brophy, P. and Clarke, Z. (2001) *EQUINOX: library performance measurement and quality management system. Deliverable D2.5: edited final report*, Version 1.0, February 2001, available at http://equinox.dcu.ie/reports/d2_5.html (accessed 6 April 2002)

Buttenfield, B. (1999) Usability evaluation of digital libraries, *Science and Technology Libraries*, **17** (3/4), 39–59.

Carter, D. S. and Janes, J. (2000) Unobtrusive data analysis of digital reference questions and service at the Internet Public Library: an exploratory study, *Library Trends*, **49** (2), 251–65.

Davidson, B. H., Salisbury, L. and Bailey, A. S. (2000) Full-text resources for undergraduates: an analysis of PA Research 2 and Wilsonselect, *College and Undergraduate Libraries*, **7** (2), 25–50.

Gorman, G. E. and Miller, R. H. (2001) Collection evaluation: new measures for a new environment. In Lynden, F. J. (ed.), *Advances in librarianship 25*, San Diego, CA: Academic Press, 67–96.

Greenstein, D. (2000) Digital libraries and their challenges, *Library Trends*, **49** (2), 290–303.

Hill, L. L. et al. (2000) Alexandria Digital Library: user evaluation studies and system design, *Journal of the American Society for Information Science*, **51** (3), 246–59.

Ingwersen, P. (1998) Web Impact Factors, *Journal of Documentation*, **54** (2), 236–43.

Jacso, P. (1998) Analyzing the journal coverage of abstracting/indexing databases at variable aggregate and analytic levels, *Library and Information Science Research*, **20** (2), 133–51.

Janes, J. (2000) Digital reference: services, attitudes and evaluation, *Internet Research* **10** (3), 256–8.

Jones, M. L. W., Gay, G. K. and Rieger, R. H. (1999) 'Project soup. Comparing Evaluations of Digital Collection Efforts, *D-Lib Magazine*, **5** (11), available at www.dlib.org/dlib/november99/11jones.html (accessed 4 July 2002)

Kantor, P. B. and Saracevic, T. (1999) Quantitative study of the value of research libraries: a foundation for the evaluation of digital libraries. In *ASIS '99. Proceedings of the 62nd ASIS Annual Meeting, Washington, DC*, vol, 36, *October 31–November 4, 1999*, Medford, NJ: Information Today, 407–19.

Kilker, J. and Gay, G. (1998) The social construction of a digital library: a case study examining implications for evaluation, *Information Technology and Libraries*, **17** (2), 60–70.

Lesk, M. (1997) *Practical digital libraries: books, bytes and bucks*, San Francisco, CA: Morgan Kaufmann.

Lipow, A. G. et al. (1998) Reference service in a digital age,. *Reference and User Services Quarterly*, **38** (1), 47.

McClure, C. R. (1999) Developing national statistics and performance measures for public libraries in the networked environment: preliminary issues for redesigning a national data collecting and reporting system in the USA. In *3rd Northumbria International Conference on Performance Measurement*, Newcastle upon Tyne: Information North for the School of Information Studies, University of Northumbria at Newcastle, 35–41.

Maquignaz, L. and O'Neil, F. (2000) Evaluation of a full text periodical database, *Lasie*, **31** (4), 66–77.

Marchionini, G. (2000) Evaluating digital libraries: a longitudinal and multifaceted view, *Library Trends*, **49** (2), 304–33.

Monopoli, M. and Nicholas, D. (2000) A user-centred approach to the evaluation of subject based information gate ways: case study SOSIG, *Aslib Proceedings*, **52** (6), 218–31.

Nisonger, T. E. (2001) The internet and collection management in academic libraries. In Liu, L. G. (ed.) *The role and impact of the internet on library and information services,* Westport, CT: Greenwood Press, 59–83.

Oppenheim, C. and Smithson, D. (1999) What is the hybrid library?, *Journal of Information Science,* **25** (2), 97–112.

Oppenheim, C. et al. (2000) The evaluation of WWW search engines, *Journal of Documentation,* **56** (2), 190–211.

Peters, T. A. (2000a) Assessing digital library services, *Library Trends,* **49** (2), 221–390.

Peters, T. A. (2000b) Current opportunities for the effective meta-assessment of online reference services, *Library Trends,* **49** (2), 334–49.

Rieh, S. Y. (2002) Judgment of information quality and cognitive authority in the web, *Journal of the American Society of Information Science,* **53** (2), 145–61.

Saracevic, T. (2000) Digital library evaluation: toward an evolution of concepts, *Library Trends,* **49** (2), 350–69.

Saracevic, T. and Covi, L. (2000) Challenges for digital library evaluation. In Kraft, D. H. (ed.), *Knowledge innovations: celebrating our heritage, designing our future. Proceedings of the 63rd ASIS Annual Meeting,* vol. 37, Silver Spring, MD: American Society for Information Science and Technology, available at www.ffzg.hr/infoz/lida/lida2000/asis_2000_text3.doc (accessed 4 July 2002)

Schwartz, C. (1998) Web search engines, *Journal of the American Society for Information Science,* **49** (11), 973–82.

Shim, W. and Kantor, P. B. (1999) Evaluation of digital libraries: a DEA approach. In *ASIS '99: Proceedings of the 62nd ASIS Annual Meeting, Washington, DC,* vol. 36, *October 31–November 4, 1999,* Medford, NJ: Information Today, 605–15.

Smith, A. G. (1997) Testing the surf: criteria for evaluating internet information resources, *Public-Access Computer Systems Review,* **8** (3), available at

http://info.lib.uh.edu/pr/v8/n3/smit8n3.html
(accessed 2 April 2002)

Smith, A. G. (1999) A tale of two web spaces: comparing sites using Web Impact Factors, *Journal of Documentation*, **55** (5), 577–92.

Smith, A. G. (2000) Search features of digital libraries, *Information Research*, **5** (3), available at www.shef.ac.uk/~is/publications/infres/paper73.html (accessed 2 April 2002)

Smith, A. G. (2001) Applying evaluation criteria to New Zealand government websites, *International Journal of Information Management*, **21** (2), 137–49.

Smith, A. G. and Thelwall, M. (2002) Web Impact Factors for Australasian universities, *Scientometrics*, **54** (3), 363–80.

Still, J. M. and Kassabian, V. (1999) Selecting full-text undergraduate periodicals databases, *EContent*, **22** (6), 57–65.

Sukovic, S. (2001) Evaluation of electronic texts, *LASIE*, **32** (2/3), 31–8.

Vogel, K. D. (1996) Integrating electronic resources into collection development policies, *Collection Management*, **21** (2), 65–76.

Wathen, C. N. and Burkell, J. (2002) Believe it or not: factors influencing credibility on the web, *JASIS: Journal of the American Society for Information Science*, **53** (2), 134–44.

13

Creating content together: an international perspective on digitization programmes

David Dawson

Introduction

There are currently a number of initiatives to share experiences in the management of digital content creation programmes. The creation of digital content is still a new area, and many aspects are untried and have yet to be evaluated. The field is rich in assertions, and poor in proven experience. In such an area the sharing of experience, and adoption of approaches that have been shown to work, can save scarce resources, both financial and human. This chapter aims to outline some of the key developments, and to identify some of the lessons that have been learned during developments in the UK.

Beginning in 2000, moves have been made within Europe to encourage shared developments, and the process is being echoed by a wider international initiative. However, these are early days, and progress depends on the detailed circumstances in each country. In the UK there have been a number of significant developments in content creation programmes over recent years. A striking feature has been the evolution from small-scale activity, largely as the result of bids by single institutions, towards large-scale funding. At the same time, the focus has been shifting towards partnerships and shared approaches by the institutions developing projects, and a more user-focused approach being taken by funders.

Cultural content projects in the UK

To begin to understand the shift within the UK, it is necessary to look well beyond the immediate institutional focus of current projects and towards the more general role that information and communication technologies (ICTs) are being expected to play in society. The British government announced its target of delivering 100% of government services electronically, initially with a target date of 2008, later brought forward to 2005. The aim of this move is to modernize government, re-align services in line with citizens' needs and to cut costs. While the government, both central and local, is investing in these services, there is a need to ensure that all citizens are able to access online services and have the training and confidence to do so. As a result, libraries are playing a critical role in ensuring public internet access to all in the UK, with the provision of £100 million from the New Opportunities Fund (NOF) to create ICT learning centres in all public libraries, £20 million to train all public librarians in the use of the internet, and £50 million to create internet-based learning resources.

Together, these funding streams are creating the People's Network. In partnership with local authorities across the UK, this is helping the process of re-engineering both the library profession and also the provision of local authority ICT provision across the UK. There are parallel streams of government funding creating other ICT learning centres, and they are all unified under the 'UK online' brand, which has received nationwide TV advertising that is linked to a telephone call centre which can tell people where to find their nearest ICT learning centre.

Museums IT Challenge Fund

It is against this background that we have seen increasing investment in the creation of cultural content, and an increasing focus on ensuring that the content has a genuine user focus. One of the earliest large-scale programmes was the Resource: the Council for Museums, Archives and Libraries, Department for Culture, Media and Sport Museums IT Challenge Fund (www.peoplesnetwork.gov.uk/content/itcf.asp). With a value of £500,000, this was open to collaborative projects among three or more museums with an objective to create materials that supported lifelong learning. A total of 11 projects were funded, and the projects were launched in Spring 2001. The projects were diverse and included 'Virtually the Ice Age', a creation of a learning resource about a prehistoric cave site at Cresswell Crags in Derbyshire, which is set in a former coalfield area and is playing an important role in the economic regeneration of the area. The website (www.creswell-crags.org.uk/virtuallytheiceage/) brings together material excavated from the site, and now held in at least four museum collections, and includes virtual reality tours of the caves themselves, most of which are not normally accessible to visitors.

Evaluation of the impact of these projects is now under way, but a key outcome of the Fund has been the development of a series of evaluation reports, commissioned by Resource. These reports include an evaluation by the projects themselves, celebrating their successes. However, shortly after the projects were completed, a series of detailed reports was commissioned under the following main headings: Skills Development, Website Creation, Partnership and Working with Communities. The aim of these evaluation projects was to identify areas that did not proceed smoothly in order to assist others in developing their own projects. As the reports cover all 11 of the IT Challenge Fund projects, it has been possible to identify 'worst practice', using a methodology that is able to focus on how to avoid the problems, and the ways found to tackle the issues, while still recognizing the great successes that the projects achieved with limited funding (www.peoplesnetwork.gov.uk/content/itcf.asp).

NOF Digitise

The People's Network, with its £170 million funding from the New Opportunities Fund (NOF), includes a £50 million programme to digitize learning resources. The NOF is a lottery funder, distributing money raised from the sale of lottery tickets to a number of good causes, under the three key themes of health, education and the environment. The programme to digitize learning resources, known as NOF-Digitise – has three main themes: cultural enrichment, re-skilling the nation and active citizenship. An initial call for proposals at the end of 1999 resulted in over 300 applications with a value of over £120 million. The projects that fulfilled the basic criteria for the NOF were invited to proceed to a second stage of application. For the first time in a major funding programme, this was undertaken with a commitment that all projects that continued to fulfil the basic criteria for the NOF would receive at least some funding. At the same time, the Fund made it clear that, in order to achieve this objective, the projects and institutions would need to be realistic about their aims and should work together to avoid duplication and to maximize the economies of scale. Central to the development of the programme has been expert advice from the People's Network Development Team, based at Resource.

While assessing the first-stage applications, it became clear to the assessment team that there were many similarities between applications and that many projects naturally clustered together on the basis either of a theme, such as shipping or migration, or geography, such as projects in Wales or the Northeast. During the second stage it was suggested that projects could work together, and particularly that they might share costs in areas such as the purchase of expensive scanning equipment, or staffing costs in areas such as project management or website design.

In order to encourage the projects to develop their plans to be as close to reality as possible, a template for a business plan was developed covering areas such as project management, audience needs assessment, content identification, website delivery and marketing. Most projects adopted this planning framework and found it a useful way to structure their planning processes. The framework has been adopted as a structure for quarterly reporting on the progress of projects, enabling projects to demonstrate achievements against milestones, and to demonstrate how they are planning to tackle the inevitable unforeseen circumstances. Developing the business plan involved a substantial amount of work by many of the applicants, and the process was aided by a support service and a series of workshops to help answer detailed questions.

Standards for digital resources

NOF-Digitise has been a ground-breaking programme, not least for the scale and breadth of the content that will be created. During the second stage of applications, a coherent set of technical standards was put in place to encourage the applicants to see their role within the wider programme, and the wider objectives of the NOF, and also to enable the development of content that would be fit for re-use and re-purposing in the future. The technical standards are based on the lifecycle model developed by Lorcan Dempsey (1999), which identified the following stages in the life of a digital resource:

- *Creation* – the actual creation of an individual digital resource.
- *Management* – the digital resource needs to be managed so that it will be accessible and meaningful.
- *Collection Development* – the digital resource typically will be placed in a collection with other digital resources. This collection will need to be managed. It should also be capable of growing.
- *Access* – materials need to be made available on the network in accessible, usable, secure and responsible ways.

- *Repackaging* – a digital resource should be able to be used in more than one way. For example, a digitized local history photograph, created initially as part of an online exhibition, may be repackaged as part of a learning resource.

As a result of this framework, the standards are intended to enable the projects to develop technical approaches which will produce materials that can be re-used in the longer term, rather than solely concentrating upon delivery of materials using current approaches. The standards were developed by UKOLN (2000), and are a distillation of best practice, and follow an extensive period of peer review. The standards are not set in stone, and are kept constantly under review. At the time of writing, Version 4 of the standards has just been made available, as a result of feedback from projects and changes in a constantly evolving field.

The technical standards cover a wide range of issues, ranging from the file formats for the digitization of images to security and data protection issues. The aim has been to present the standards as guidelines, and using the Internet Engineering Task Force definitions:

> The words '*must, should* and *may*' when printed in bold text have precise meanings in the context of this document.
>
> *Must*: This word indicates an absolute technical requirement with which all projects must comply.
>
> *Should*: This word indicates that there may be valid reasons not to treat this point of guidance as an absolute requirement, but the full implications should be understood and the case carefully weighed before it is disregarded. *Should* has been used in conjunction with technical standards that are likely to become widely implemented during the lifetime of the project but currently are still gaining widespread use.
>
> *May*: This word indicates that the topic deserves attention, but projects are not bound by New Opportunities Fund advice. *May* has therefore been used to refer to standards that are currently still being developed.
>
> (UKOLN, 2002)

This enables a clear distinction to be made between advice and mandatory requirements, as illustrated by the following extract dealing with the creation of digital images.

> Images *must* be created using one of the following formats – TIFF, PNG, GIF or JPEG/SPIFF. In general photographic images *should* be created as TIFF images. However, in cases, for example when using cheaper digital cameras, it may be appropriate to use JPEG/SPIFF as an alternative. This will result in smaller, but lower quality images. Such images may be appropriate for displaying photographs of events etc. on a Web site but it is not suggested that such cameras are used for the large-scale digitisation of content. Line drawings, such as certain types of computer-generated imagery, *should* normally be created as PNG or GIF images.   (UKOLN, 2002)

By contrast, the following refers to the delivery of images over the internet:

> Images *must* be provided on the Web as GIF (for line-drawings) or JPEG/SPIFF (for photographs) formats. PNG *may* be an alternative format to GIF.   (UKOLN, 2002)

It is important to stress, though, that these technical standards are in no way intended as a straitjacket on innovation, or as encouraging the development of a bland set of projects conforming to some overly prescriptive standardized world view. Rather, the standards build upon well-established best practice, and lay a set of stable, interoperable, sustainable foundations upon which truly innovative and worthwhile projects will be built and maintained. The advantages of this approach have been clearly demonstrated by the reaction of the projects that have received funding and also by the fact that similar approaches are being taken in other programmes – the recent *Standards and guidelines for digitization projects* (Alexander and Kuny, 2001) for Canada's Digital Cultural Content Initiative (www.pch.gc.ca/cdcci-icccn/) and *Working with the DNER: standards and guidelines to build a national resource* (JISC, 2001), for example, are both closely based on the NOF guidelines, and there is further interest in the USA and elsewhere to similarly build upon this foundation.

The NOF-Digitise technical standards also aim to ensure that the websites developed by the projects support universal design – ensuring accessibility for the visually impaired and cross-platform delivery, ready for the future deployment of 3G mobile communications technologies. The principles developed in these technical standards anticipated the development of e-government standards by the British government – the e-Government Interoperability Framework (e-GIF). While the e-GIF was developed by central government, the principles in fact apply across the whole public sector. The current definition of this is any organization that is listed in the *Civil Service yearbook*, including all local authorities, national museums and libraries, and also major Lottery funders. As a result, these standards apply to the projects being funded by the NOF, and the projects are well placed to take a lead in the adoption of the e-GIF. Many of the projects have been developed by museums, libraries or archives that form part of a local authority and as a result, a number of these local authorities are in the process of re-designing their websites to meet e-GIF requirements, including World Wide Web Consortium (W3C) criteria to ensure accessibility for the visually impaired.

Building on experience

The NOF-Digitise programme has been fortunate in being able to build upon the extensive experience of the £15 million Electronic Library (eLib) programme of the Joint Information Systems Committee (JISC). JISC is funded by a top-slice funding from the higher and further education sectors of the UK and includes the funding of the SUPERJanet high-speed network, as well as the provision of a wide range of electronic content and learning materials. In 1995 JISC launched the eLib programme, consisting of over 160 projects, which resulted in the creation of a number of services that under-pin much of the content creation and delivery for the higher and further education sector. Outstanding examples include the UK Office for Information and Library Networking (UKOLN), which has been driving much of the standards adoption work, both in the UK and elsewhere, and the Arts and Humanities Data Service (AHDS). The AHDS maintains access to, and archives data from, a number of arts- and humanities-based subjects and themes, including history, performing arts, visual arts, electronic texts and archaeology. These organizations, together with others, such as the Higher Education Data Service and the Technical Advisory Service for Images, have created an extensive body of knowledge about the digitization and delivery of online content. This wealth of experience has been available to the NOF-Digitise projects through a combination of a contracted advisory service, *pro bono* advice and consultancy services for individual projects.

NOF-Digitise projects

The projects that have been successful in receiving funding from the NOF cover a bewildering range, with 152 different projects involving over 500 different institutions. The projects range from community information to the voluntary sector, health, skills development and the cultural sector. The programme will create over 1 million digital objects – including text, images, audio and video clips, and will also include over 400 learning packages. It is difficult to give an overview of the range of projects, but three examples illustrate their range and diversity.

SEAMLESS citizens' access to digitized sources – Essex Libraries
www.seamless.org.uk

Essex Libraries are developing a web-based Citizens' Gateway, a single point of access to integrate information from key national and local suppliers of information. The Gateway will provide access to a wide variety of community and citizenship information at local, regional and national level, depending on the particular subject. Information will come from a variety of sources. The main subject areas include health and welfare, legal and citizens' rights information, education and lifelong learning, leisure and employment.

British Pathe Film Archive
www.britishpathe.com

The main aim of this project is to make all 50,000 'stories' and 15,000 unpublished items within the archive available online for viewing in schools and libraries. To achieve this the archive, currently held on 35mm film and available only to media professionals, will need to be re-mastered, re-catalogued and digitized to the highest possible specification. From 1910 to 1970 British Pathe produced a bi-weekly newsreel, which covered politics, domestic issues, sport and international affairs for a huge cinema audience. From 1918 to 1969 British Pathe also produced the famous weekly 'Pictorials', which charted cultural activities in Britain such as fashion, art and music.

Voluntary Matters – the Media Trust
www.voluntarymatters3.org

The Voluntary Matters Project will convert into digital form the training material contained in or pertaining to the first two series of 'Voluntary Matters' videos and texts, originally broadcast by the BBC, to provide engaging, interactive, web-based training for the voluntary sector. When shown on the BBC's Learning Zone, 'Voluntary Matters' attracted more than 20,000 viewers per programme. The two series were entitled 'A guide to management and good practice in the voluntary sector' and 'Building your future in the voluntary sector'.

All the 150 projects will have at least some presence on the web by the end of 2002 and will be completed during 2004. At the time of writing, there is a website (www.nof-digitise.org) that lists all the projects, and there are plans to create a portal that will encourage users to explore a wide range of materials, and to search across projects. While the technical details that will be adopted to deliver this service are still being finalized, the People's Network is working with a number of other key players, including JISC, the National Health Service and the British Library, to investigate the potential shared development of a Common Information Environment. JISC is developing a service, the Distributed National Electronic Resource (www.jisc.ac.uk/dner), which will be using a range of technologies, including authentication services, resource discovery and cross-searching to create an information environment for the higher and further education sectors, integrating public and commercial information services.

Culture Online

The experience, ideas and enthusiasm of the NOF-Digitise programme are being taken forward in a new initiative developed by the Department for Culture, Media and Sport. Initially proposed by the then Culture Minister, Chris Smith, in September 2000, a report (Great Britain. DCMS) and website (www.cultureonline.org.uk) were prepared to illustrate the vision. The executive summary emphasizes how the proposal aims to provide the materials and resources that people want – user driven rather than producer led.

> Culture Online's aim is to deliver the riches and know-how of our museums and galleries, libraries and archives, theatre companies and orchestras, into classrooms and living rooms across the country to enrich opportunities for learning for people of all ages.
>
> By creating a digital bridge between culture and learning we would make both sides of that relationship more productive. We could make more productive use of the wealth of publicly funded cultural capital –

the archives, collections and know-how of curators, artists and directors – by creating much wider access. School children would have access to the teaching of experts in their field. Many of our museums and galleries are pioneering work with Web sites. However, many are well behind the leaders. Without more concerted investment we are likely only to scratch the surface of the opportunity to make the assets of the cultural sector available for learning.

The market is unlikely to grasp this opportunity; that is why there is a case for public intervention. Venture capitalists are retreating from investment in new Internet start-ups. Although large media companies have plans for E-learning, these are unlikely to make more than passing use of the expertise built up in our publicly funded cultural sector.

A new body, working in partnership with cultural organisations, would lend greater momentum to their efforts, by creating a network of resources, linking all our cultural institutions, often in collaborative projects, to pool ideas and know-how. Without a body to promote collaboration and spread best practice there is a danger we will see a scattering of fragmented and isolated initiatives.

(Charles Leadbeater in Great Britain. DCMS)

The Department has submitted a detailed business plan for the project to the Treasury, and the outcome is eagerly awaited. Culture Online is seen as an important project, not just for the cultural sector, but also as part of the wider government objective to promote the adoption of broadband technology in the UK. The roll-out of broadband has been slow in Britain, and a two-pronged approach is required: to encourage the telecommunications industry to invest in new broadband networks, and to ensure that content is created that will encourage consumers to subscribe to the new broadband services. The UK Broadband Stakeholders Group, led by the Office of the e-Envoy, is currently developing strategy in this area. There are great similarities with the situation in other European countries, as well as in the USA and Canada.

Co-ordinating European digitization policies

In Europe the e-Europe Action Plan (European Commission, 2000) is promoting the adoption of new technology, in order to help build a European Information Society. The Action Plan was adopted by all the member states in Europe at the Feira European Council in June 2000, and proposed ten priority areas for action with ambitious targets to be achieved through joint action by the Commission, the member states, industry and the citizens of Europe. These areas of action are:

1 *European youth into the digital age*: bring internet and multimedia tools to schools and adapt education to the digital age.
2 *Cheaper internet access*: increase competition to reduce prices and boost consumer choice.
3 *Accelerating e-commerce*: speed up implementation of the legal framework and expand use of e-procurement.
4 *Fast internet for researchers and students*: ensure high-speed access to internet thereby facilitating co-operative learning and working.
5 *Smart cards for electronic access*: facilitate the establishment of European-wide infrastructure to maximize uptake.
6 *Risk capital for high-tech SMEs*: develop innovative approaches to maximize the availability of risk capital for high-tech SMEs (small and medium-sized enterprises).
7 *'E-Participation' for the disabled*: ensure that the development of the information society takes full account of the needs of disabled people.
8 *Healthcare online*: maximize the use of networking and smart technologies for health monitoring, information access and healthcare.
9 *Intelligent transport*: provide safer, more efficient transport through the use of digital technologies.
10 *Government online*: ensure that citizens have easy access to government information, services and decision-making procedures online (European Commission, 2000).

One of the detailed actions proposed was to co-ordinate national digitization policies, recognizing especially the critical roles for cultural
content creation in:

- preserving Europe's collective cultural heritage
- providing improved access for citizens to that heritage
- enhancing education and tourism
- developing e-content industries.

A meeting of representatives and experts from the member states
gathered in Lund and identified ways in which a co-ordinating mechanism for digitization programmes across the member states could be
put in place to stimulate European content in global networks, in
order to meet Objective 3(d) of the e-Europe Action Plan (European
Commission, 2000). The meeting also identified many obstacles to
the success of these initiatives. These hurdles include the diversity of
approaches to digitization, the risks associated with the use of inappropriate technologies and inadequate standards, the challenges
posed by long-term preservation and access to digital objects, lack of
consistency in approaches to intellectual property rights, and the lack
of synergy between cultural and new technology programmes.

The Lund Meeting concluded that these obstacles could be tackled
and the objectives of the e-Europe Action Plan could be advanced if
the member states were to

- establish an ongoing forum for co-ordination
- support the development of a European view on policies and programmes
- develop mechanisms to promote good practice and consistency of
  practice and skills development (European Commission, 2001).

An initial survey of content creation policies, programmes and projects was undertaken, and the results indicated that this was a useful
activity, but that more work was required, particularly on the survey-

ing of projects. Key issues here are the lack of a co-ordinating body in many countries, and therefore the difficulty of recording information about projects funded by institutions themselves, as well as by a range of private bodies, local and national government and through collaboration with industry.

As a result, some initial work on a metadata framework for project resource discovery was begun, centring on the investigation of two complementary approaches:

- the development of a metadata schema by the Ministry of Culture in France, used to record nationally funded projects
- the Collection Description Framework developed for the UK Research Support Libraries Programme (www.rlsp.ac.uk) and now carried forward by the Collection Description Focus (www.ukoln.ac.uk/cd-focus).

At the time of the London meeting it was anticipated that additional work in this area would be undertaken in Autumn 2001, and that this might prove a valuable framework for discussion at the Washington meeting.

In the event, further work in this area will be undertaken under the leadership of the French Ministry of Culture, in the Minerva Project, funded by the European Commission. The Minerva Project, hosted by Italy and also including partners in the UK, Sweden, Spain, Finland and Belgium, is a mechanism for tackling the issues identified in the Lund Action Plan.

The objectives for this work are:

- increasing the visibility and accessibility of European cultural and scientific content by setting up inventories of ongoing digitization projects based on national observatories
- definition of a sustainable technical infrastructure for co-ordinated discovery of European digitized cultural and scientific content
- analysis of possible solutions to the technical and cultural con-

straints deriving from multilingual problems
- proposal of a common platform (XML and open source) for accessing distributed information in Europe (www.amitie.it/minerva/wp/wp3.htm).

International collaboration

These issues being tackled within Europe are also being faced on an international basis. Recognizing the common issues, a small group of decision makers and policy developers from Europe, Canada, the USA and New Zealand met in London in June 2001, to explore the extent to which each could learn from the lessons of the others, and to identify the areas in which collaborative work might be of mutual benefit.

The incentive for the meeting held in London was a recognition that, although an increasing amount of work was being done, it tended to take place on a national basis, such as Canada's *Standards and guidelines for digitization projects* (Alexander and Kuny, 2001), at a pan-national level, such as the Lund Principles (European Commission, 2001) or by individual projects or consortia – such as National Initiative for a Networked Cultural Heritage (NINCH)'s Best Practice Working Group (NINCH, 2002) or the guidelines developed by RLG for their Cultural Materials Initiative (Gill, 2002).

Although all of these were, and continue to be, important in advancing the situation, none appear to have the breadth, scope, and mandate to tackle a range of problems internationally. Thus the June 2001 meeting, at which the attendees explored the scope for putting together much of the existing work and producing outputs explicitly aimed beyond the relatively narrow constituencies of the existing efforts.

Over two days of meeting in London, participants covered a wide range of topics. There was clear consensus on the need to improve the current situation, and on areas in which work should most urgently be taken forward. Included in the list of actions was the drafting of position statements that seek to identify key issues and to provide a strong policy support on the following issues:

- the relevance and value of cultural content creation to society
- the importance and urgency of problems around digital preservation
- the issues surrounding the protection and exploitation of IPR in the networked environment (Miller, Dawson and Perkins, 2001a).

The group also identified a number of common areas where early information sharing will either inform the development of further position statements or indicate the urgent need for a co-ordinated approach to research. The following areas were identified at this initial stage:

- usage surveys, market analysis, impact research and identification of latent demand
- economic and business factors informing sustainability models
- national content creation policies to identify common pan-national issues
- objectives, criteria and results of major content creation programmes
- objectives, criteria and results of relevant major technical research programmes
- gathering existing technical standards, policy documents, best practice guidelines to identify the potential for harmonization (Miller, Dawson and Perkins, 2001a).

Following the meeting, an article was published (Miller, Dawson and Perkins, 2001b) which ensured wider discussion of the issues raised, and gained the attention of a wider audience. At a second meeting, held in Washington in March 2002, more countries and a wider range of stakeholders were involved and agreed to continue working together, placing a special emphasis on the need to conduct research, drawing upon the common lessons of existing research undertaken on user needs and the ways in which users interact with cultural materials.

Conclusion

The key lesson learned so far is that user needs must be at the heart of digitization projects. A user who is interested in finding out about the history of their family or town does not care whether the information comes from a local organization or an organization based on the other side of the world – they simply need to know that the information is accurate and reliable. It is meeting this challenge that will require new ways of working, new collaborations, and a shared view of the technical solutions needed to deliver the user-focused services of the future. The foundations for this infrastructure are being planned now. We now need to make sure that the architect's brief will meet the needs of the client.

References

Alexander, M. and Kuny, T. (2001) *Standards and guidelines for digitization projects, Canadian Digital Cultural Content Initiative*, available at
www.pch.gc.ca/ccop-pcce/pubs/ccop-pcceguide_e.pdf
(accessed 15 May 2002)

Dempsey, L. (1999) Scientific, industrial and cultural heritage: a shared approach, a research framework for digital libraries, museums and archives, *Ariadne*, **22**, available at
www.ariadne.ac.uk/issue22/dempsey/
(accessed 15 May 2002)

European Commission (2000) *e-Europe action plan*, available at
http://europa.eu.int/information_society/eeurope/action_plan/
actionplantext/index_en.htm
(accessed 15 May 2002)

European Commission (2001) *Lund Principles: conclusions of experts meeting, European Commission*, available at www.cordis.lu/ist/ka3/digicult/lund_p_browse.htm (accessed 15 May 2002)

Gill, T. (ed.) (2002) *RLG cultural materials alliance description guidelines*, available at www.rlg.org/culturalres/descguide.html (accessed 15 May 2002)

Great Britain. Department for Culture, Media and Sport *Culture Online* www.cultureonline.gov.uk/vision.pdf (accessed 15 May 2002)

Joint Information Systems Committee (2001) *Working with the DNER: standards and guidelines to build a national resource*, available at www.jisc.ac.uk/dner/programmes/guidance/DNERStandards. html (accessed 15 May 2002)

Miller, P., Dawson, D. and Perkins, J. (2001a) *An international seminar on national digital cultural content creation strategies*, available at www.ukoln.ac.uk/interop-focus/activities/events/digitisation-strategies/outcomes (accessed 15 May 2002)

Miller, P., Dawson, D. and Perkins, J. (2001b) Standing on the shoulders of giants: efforts to leverage existing synergies in digital cultural content creation programmes world-wide, *Cultivate Interactive*, **5**, available at www.cultivate-int.org/issue5/giants (accessed 15 May 2002)

NINCH (2002) Guide to good practice in the digital representation and management of cultural heritage materials, NINCH Best Practices Working Group, described at www.ninch.org/programs/practice

UK Office for Information and Library Networking (2000) *NOF-*

Digitise technical standards, version 1, available at
www.peoplesnetwork.gov.uk/content/technical.asp
(accessed 15 May 2002)
UK Office for Information and Library Networking (2002) *NOF-
Digitise technical standards, version 4*, available at
www.peoplesnetwork.gov.uk/content/technical.asp
(accessed 15 May 2002)

Part 6

STANDARDS AND TECHNOLOGY

14

Making sense of metadata: reading the words on a spinning top

Daniel G. Dorner

Introduction

This chapter owes a debt to Brandon Hall for his insight that making 'sense out of standards is as difficult as trying to read words on a spinning top' (Hall, 2001). Capturing and making sense of the rapidly changing literature on metadata is almost as difficult. Less than ten years ago, the term 'metadata' was practically unknown to librarians and information managers. Today, it is one of the hottest topics in our literature.

The term entered our vocabulary concomitantly with the rapid growth in the number of digital information resources and more specifically with the adoption of the world wide web as the primary means for distributing those resources. Now metadata is the focus of numerous journal articles, workshops, conferences and international meetings involving librarians, archivists, museum curators, information systems managers, computer scientists and other information-related professionals.

This chapter provides a synthesis of some of the key issues and developments related to metadata over the past four years, especially with reference to metadata for resource description and discovery.

The chapter does not attempt to be a guide to how to create meta-data or how to establish a metadata project. However, the aim is to help librarians and information managers, especially those who are coming to grips with the stark reality of managing electronic infor-mation resources, become better informed about metadata and its roles, about the variety of metadata standards being developed and implemented, and their potential for enhancing access for specific communities to their digital information collections.

The rapid rise of metadata

The growing importance of metadata in the library and information management literature is easily illustrated through the results of searches on one of the discipline's pre-eminent online databases – Library and Information Science Abstracts (LISA). A keyword search on LISA for the term 'metadata' or the terms 'meta' and 'data' retrieved a set of 847 entries. Figure 14.1 shows the dramatic increase in the use of the term that began around the middle of the 1990s.

A good starting point to read about metadata is Vellucci's 1998 chapter in the *Annual review of information science and technology* sim-ply entitled 'Metadata' (Vellucci, 1998). The book *Getting mileage out of metadata* by Hudgins, Agnew and Brown (1999) is another excellent introductory source. It aims to be 'a starting point for libraries of all types that are planning for the catalog of the future by incorporating metadata standards into their cataloging repertoire'. However, each of these works predates most of the literature on the topic.

The term 'metadata' appears to have surfaced in the literature for the first time in 1978. At the 1978 Annual Conference of the American Society of Information Science, Cousins and Dominick (1978) discussed the concept of a 'meta-data base (or meta-base)' as a tool for managing a multiplicity of databases. They presented the case for developing a resource that would contain 'the structural and semantic data of other bases and information systems designed to support them' and they discussed 'the types of information which a

meta-base must contain'. In their paper they also considered 'user language design, vocabulary control and searching within systems
designed to support a meta-base philosophy'.

Four years passed before a special issue of the *Drexel Library
Quarterly* on numeric databases included two articles in which the
term metadata appeared in their abstracts. In the abstract for one of
these articles Liston and Dolby (1982) defined metadata as 'information about numeric data', whereas in the abstract for the other article, Widerkehr (1982) focused 'on a particular kind of metadata,
namely data element linkage information . . . including backward and
forward tracing'. In the same year in the conference proceedings of
the American Society for Information Science Lipkin (1982) discussed 'how the problems and challenges encountered in developing
an automated metadata system differ from the more familiar ones
associated with automated data system development' in the context
of data resource directories.

The term 'metadata', however, almost dropped completely out of
the professional vocabulary for a dozen years, appearing in only one
paper indexed in LISA in 1987, one in 1990, two in 1992 and three in
1993, before starting on its meteoric rise in use in 1994.

Figure 14.1 shows the dramatic rise of use of the term 'metadata'
in the literature from 1994 and 1999. New entries for articles on

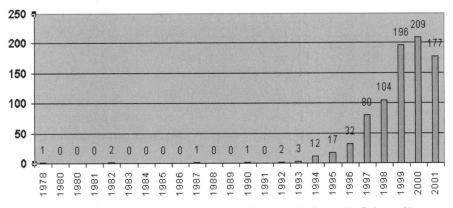

Fig. 14.1 *Number of metadata articles by year in Library and Information Science Abstracts*

metadata will likely be added to LISA for the year 2001, so the number of entries for that year will probably be similar to 2000 and 1999, suggesting a plateau has been reached, at least for the time being. Given the explosion of material on metadata in the latter half of the 1990s, an obvious question is: why did metadata rise so dramatically in importance from 1994 to 1999? The following section examines explanations provided within the literature.

The growing importance of metadata

As mentioned earlier, the rapid rise of metadata in the literature closely mirrors the adoption in the mid-1990s of the world wide web (WWW) as the primary means for distributing electronic information resources. Pottenger, Callahan and Padgett (2001) comment that 'metadata has become more important than ever before, because it can facilitate the organization and management of networked information'.

Although Lange and Winkler (1997) point out that the term 'metadata' began to appear in the database management systems literature in the 1980s – prior to its rise in the LIM literature – Milstead and Feldman (1999) argue that librarians and indexers have been producing and standardizing metadata for centuries without actually knowing it. Their view is supported by writers such as Vernon and Rieger (2002),who contend that the history of metadata can be traced to library catalogues which have long provided 'both intellectual and physical access to a collection'.

Vellucci (1998) maintains that, whereas the library science community traditionally used the terms 'bibliographic data or cataloging data', in the electronic environment the methods used to organize information in both the library science and information science communities have converged and the two communities have adopted the term 'metadata' for describing electronic data. Vellucci also describes some of the tensions that have arisen within the bibliographic community related to the adoption of 'metadata' as a term. She refers to

Larsgaard (1996), as cited in Vellucci (1998), who considers it ironic that what is essentially cataloguing requires a non-cataloguing term in order to be accepted by non-cataloguers. She also comments on Caplan's perspective (see Caplan, 1995) that the term 'catalogue' has negative connotations, so the term 'metadata' has been adopted because it offers neutrality for those outside the cataloguing environment.

Michael Gorman (1999) argues that the non-cataloguing world perceives metadata as being different from traditional cataloguing, which is believed to have complex formats and expensive and stringent quality requirements. From the perspective of the non-cataloguers, according to Gorman, metadata provides 'a third way' fitting between traditional cataloguing and the other extreme, i.e. the reliance on key-words, an unstructured system that is largely ineffective but also very inexpensive. After Gorman describes the 15 elements of the most widely adopted metadata standard, the Dublin Core, he says 'it is easy to see that what we have here is, in essence, a sub-set of MARC' (Gorman, 1999). Gorman then points out that the 'minimalist view' behind Dublin Core is that its simplicity is its strength. He argues, however, that

> it really boils down to a choice between an inexpensive and ineffective form of cataloguing in which the 15 elements of the Dublin Core are filled with unqualified and uncontrolled free text on the one hand or an expensive and effective form of cataloguing in which at least some of the elements of the Dublin Core are filled with normalized controlled data decided on the basis of professional examination of the resource.   (Gorman, 1999)

Gorman believes that four approaches are available for the bibliographic control of electronic resources of varying quality: full MARC cataloguing for high-quality resources that are likely to have continuing value, enriched Dublin Core 'cataloguing' records for the next level, minimal Dublin Core records for the next, and relying on unstructured full-text keyword searching for the remainder (Gorman, 1999). In other words he believes that metadata standards such as the Dublin Core do not provide adequate bibliographic control for the kinds of electronic resources that will have enduring value for library collections and their users, and that we need to rely on MARC and the *Anglo-American cataloguing rules* (2nd edn) to provide access to those resources.

Other writers, however, take a somewhat different perspective without necessarily disagreeing with Gorman. Medeiros, for example, points out that with the rise of web portals and search engines, the library catalogue is no longer the primary research tool for many library users, although it will continue to play an important role in providing access to information (Medeiros, 1999). He suggests that technical services librarians are burdened with principles based on the print model, and to remain relevant they need to embrace the emerging metadata standards. Brisson (1999) contends that in visualizing the library's role in the digital era, librarians must be mindful of its traditional functions but they also must be cautious about carrying over their traditional activities in a wholesale manner to the digital library. He argues that the concepts of the digital library and metadata 'hold the potential for integrating the essential functions of the library into the digital environment, and strategically centering the library for the critical role it should play in the coming digital society of the 21st century' (Brisson, 1999). Brisson and other authors perceive metadata as providing a middle way for libraries to provide quality access to the large numbers of digital resources that will make up the digital libraries of the future.

Defining 'metadata'

The question arises, then, if metadata is not traditional cataloguing for electronic resources, what is it? Definitions of metadata abound, the most common of which is 'metadata is data about data'. However, the simplicity of this definition renders it almost meaningless, given the richness that the concept is afforded when it is examined in more detail.

Gradmann (1998) argues that cataloguing data and metadata differ on several counts. The former is created for library users by trained cataloguers, whereas the latter is created for the use of software agents by the authors of the documents. He also contends that cataloguing data provide resources descriptions whereas metadata is meant for resource discovery.

Gilliland-Swetland (2000) points out that 'metadata' is now becoming an increasingly ubiquitous term but it is being understood in different ways by different professional communities. She says that, until the mid-1990s, the communities involved in managing geospatial data, and in designing and maintaining data management systems, used 'metadata' to refer to 'a suite of industry or disciplinary standards' and to a range of documentation and data required 'for the identification, representation, interoperability, technical management, performance, and use of data contained in an information system' (Gilliland-Swetland, 2000). However, she argues that it is more useful to see the big picture and think about metadata as

> the sum total of what one can say about any *information object* at any level of aggregation. In this context, an information object is anything that can be addressed and manipulated by a human or a system as a discrete entity. The object may be comprised of a single item, or it may be an aggregate of many items. In general all information objects, regardless of the physical or intellectual form they take, have three features – content, context, and structure – all of which can be reflected through metadata.   (Gilliland-Swetland, 2000)

Another useful definition is the one adopted by the Association for Library Collections and Technical Services' (ALCTS) Task Force on Metadata. Among other things, the Task Force was charged with devising a definition of 'metadata' and investigating the interoperability of newly emerging metadata schemes with the cataloguing rules and MARC format (ALCTS Task Force on Metadata, 2000). The members of the Task Force decided to provide definitions for the related terms of 'metadata', 'interoperability' and 'metadata schema'. However, they found the work to be more difficult than anticipated because of the varying intents and contexts that can surround the definitions. Based on the majority perspective, the following working definitions were provided in the final report of the Task Force:

> Metadata are structured, encoded data that describe characteristics of information-bearing entities to aid in the identification, discovery, assessment, and management of the described entities.
>
> Interoperability is the ability of two or more systems or components to exchange information and use the exchanged information without special effort on either system.
>
> A metadata schema provides a formal structure designed to identify the knowledge structure of a given discipline and to link that structure to the information of the discipline through the creation of an information system that will assist the identification, discovery and use of information within that discipline.
>
> (ALCTS Task Force on Metadata, 2000)

Fietzer (1999) contends that the most valuable lesson from the Task Force's work is its recognition of the importance of the context in which the metadata is employed. He says 'metadata is fluid, mutable, and dynamic', and it mirrors the data it describes by changing over time, and doing different things and different times. The key factor of metadata is that 'the things it does reflect the needs and interests of the community or communities using it' (Fietzer, 1999).

The Task Force's definition fits very closely with Gilliland-Swetland's definition discussed above. A striking feature of both definitions is that rather than referring specifically to electronic information resources, they are format-free so to speak, referring to 'information objects' (Gilliland-Swetland, 2000) and 'information-bearing entities' (ALCTS Task Force on Metadata, 2000). On the other hand the focus of the descriptive rules in traditional cataloguing (i.e. AACR2) is on specific formats. This specificity is one of the five major problems raised by Taylor (1999) with using AACR2 for describing internet resources. Metadata, therefore, is applicable across all formats of information, though the term is used predominantly in relation to electronic resources, regardless of whether the resources are: standalone or networked, simple (e.g. a single web page) or complex (e.g. a web page with many links), tangible (e.g. a CD-ROM) or intangible (e.g. a web page). Day (2001) says that 'the general understanding of the term has since broadened to include any kind of standardised descriptive information about resources, including non-digital ones'.

The above definitions also highlight the fact that metadata is used for more than simply describing networked information resources, and the contextual aspects that the definitions raise provide a useful framework for analysing metadata further.

From the above definition we can see that metadata can play a variety of roles in support of a wide range of operations, including resource identification and the management of rights associated with those resources, resource description and discovery, the management of information resources within institutions and their interoperability across institutions and systems, and the long-term preservation of information resources. It is the first two of these roles on which the remainder of this paper focuses.

Resource identification and rights management

One of the great problems of networked electronic information resources such as websites is that they are often transient in nature. According to Davidson and Douglas (1998), the transience of websites, 'coupled with the difficulty of discovering and retrieving their contents, causes serious problems for libraries and other institutions that attempt to integrate Web documents into local systems'. In addition electronic publishers have been concerned with the effortless nature of cutting and pasting all or part of electronic resources into new resources, or of copying and replicating digital material with or without authorization to large numbers of other people (Davidson and Douglas, 1998). These concerns have led various groups to develop systems for providing persistent identifiers.

In a report to the National Library of Australia Dack (2001) explains that the stability of links between digital resources is one of the keys to the successful development of distributed digital libraries or archives. She describes both general and specific formal identifier schemes and the issues and problems associated with them. Her discussion of general schemes focuses on Uniform Resource Identifier (URI) systems (such as the Universal Resource Locator) and Universal Resource Names (URNs). Her discussion of more specific naming schemes includes detailed description of the syntax, as well as other issues and concerns pertaining to the following schemes for which URLs have been added to assist in locating more information:

- the National Bibliography Number (NBN) developed by the National Library of Finland
  http://rfc.sunsite.dk/rfc/rfc3188.html
- the Persistent URLs developed by OCLC
  http://purl.oclc.org/
- the Handle system developed by the Corporation for National Research Initiatives
  www.handle.net/
- the Digital Object Identifier (DOI) system developed by the American Association of Publishers (see discussion below)
  www.doi.org/
- the Archival Resource Key (ARK) developed by John Kunze for the US National Library of Medicine
  www.ckm.ucsf.edu/people/jak/home/

An important feature of Dack's report is her identification of the types of electronic resources on the web that need to be managed for persistence. She says, 'The best rule of thumb is that any item which has been made publicly available online and which is likely to have been cited or referenced in other digital objects or online resources should be organised in such a way that citations to it persist' (Dack, 2001). She divides these resources into 'static resources', i.e. those with a definitive version of the resource that won't be altered, and 'dynamic resources', i.e. those that undergo change by modification, deletion or insertion of content, which can be either ephemeral or in need of archiving. Dack observes that the level and type of persistent identifiers required will be dependent on the nature of the material itself and the requirements of the organization managing the resource.

> A detailed analysis and typology of the information on the site and policies on the identification and organisation of the material will enable those responsible for managing it to decide at the point of creation of a document what level of persistence is required and it can be located, flagged and named accordingly.   (Dack, 2001)

A variety of organizations, including national and state libraries and archives, institutions involved in creating and managing distributed digital collections, and others concerned with intellectual property rights, have an interest in developing persistent identifier systems to meet the needs of their communities.

Digital Object Identifiers

The Association of American Publishers (AAP) provides an example of an organization that has analysed its members' needs carefully and has developed a persistent identifier system to meet those needs. In 1994 intellectual property concerns led the AAP to develop a system based on the creation of the Digital Object Identifier (DOI) (Davidson and Douglas, 1998). The DOI is a 'wholly unique number used to unambiguously label a piece of intellectual property', which 'follows a structure similar to that of the common Web URL and other Uniform Resource Identifiers (URI)' (Miller, 2000a).

The system has now become international in scope. In 1998 the International DOI Foundation (IDF) was established to support 'the needs of the intellectual property community in the digital environment, by the development and promotion of the Digital Object Identifier system as a common infrastructure for content management' (Paskin, n.d.). Miller points out that, by associating the DOI with other metadata, it is possible to provide 'additional contextual information of use both to human users and to automated tools acting on the user's behalf' (Miller, 2000a). The great advantage for libraries and people searching the web is that DOIs resolve much of the transience associated with digital information resources. According to Miller,

> When a document is assigned a DOI by a publisher, it is the responsibility of the publisher to lodge the DOI, the URL to which it points, and associated metadata with the resolution service. A user entering a DOI will have it automatically resolved, and will be pointed to the URL at which the document can be found. If the URL changes for some reason, the entry in the resolution database is simply changed and the DOI remains unaltered, ensuring a degree of permanence to the underlying intellectual content which the user is looking for.
>
> (Miller, 2000a)

The first application of the DOI has been a linking system called CrossRef which 'was incorporated in January of 2000 as a collaborative venture among 12 of the world's top scientific and scholarly publishers' (Brand, 2001). As of May 2002, 'there are 122 publishers participating in CrossRef, accounting for over 6300 journals with over 4.8 million article records in the database' (Publishers International Linking Association, 2002). According to Brand, CrossRef benefits both scholarly writers, who can link automatically to cited electronic documents, and publishers, whose publications increase in demand and use, and thus in value to their users (Brand, 2001).

Metadata for users

An associated and very interesting aspect of the rights management issue relates to the suggestion that metadata describing individual users can help meet the intellectual property protection requirements of the various players in the information chain. In a report for the UK Library and Information Commission and the British National Bibliographic Research Fund, Potter and Bide (1999) examine the management by the academic library and commercial user communities of access to electronic academic journals in terms of user identification, authentication and authorization. Concluding that the current methods fall short of the requirements, they propose a system in which the users' metadata is embodied in user 'passports' and 'visas':

> The model proposes that each individual possesses a User Passport which identifies both who one is and something about the groups and classes to which one belongs. Within the passport, Visas exist which indicate that the individual has certain rights in respect of certain resources. Some of these rights may derive from institutional membership; others might be acquired via an individually negotiated arrangement.   (Potter and Bide, 1999)

A 'passport' that incorporates user metadata should make it easier for the user to access information resources for which he or she is authorized, and should also allow authentication of users when access requires a commercial transaction. We need to be aware, however, that Microsoft is also using the passport concept by issuing '.NET Passports' for controlling access and tracking purchases of its products. This has led to concerns in some quarters about privacy issues (see, for example, Mainelli 2002).

In essence, identification metadata for electronic information resources assists libraries and individuals to retrieve those resources. It also helps publishers manage and promote those resources. On the other side of the ledger, the identification metadata for users helps the publishers of electronic information resources and the libraries that have purchased licences for those resources to manage access to those resources. For individual users the identification metadata will provide them with authentication as bona fide users, and, it is hoped, with simplified access to the electronic resources.

Resource description and discovery

One of the primary purposes of metadata is to support resource description and discovery. The Dublin Core Metadata Initiative, for example, traces its history back to a 1995 meeting in Dublin, Ohio in which more than 50 people from diverse information backgrounds 'discussed how a core set of semantics for Web-based resources would be extremely useful for categorizing the Web for easier search and retrieval' (Dublin Core Metadata Initiative, 2000).

Metadata for resource description and discovery establishes the content of an electronic resource and is intrinsic to that resource (Gilliland-Swetland, 2000). These intrinsic qualities include characteristics such as the intellectual content, the creator and the physical form of the resource. This content metadata is, therefore, essential for organizing and retrieving electronic information resources and typically is developed by and for a specific community of interest.

As the importance of metadata has grown in tandem with the rise in number of networked electronic information resources, so too have the number and types of communities involved in establishing metadata for describing and discovering those resources. These communities range from those that are subject specific or institution specific (see Bakewell et al., 1999), to those that are involved in disciplinary or commercial areas (see Education Network Australia, 2000, and Association of American Publishers 2000), to those that are more general in scope (see Hillman, 2001). However, this diversity also means that there are increasing numbers of metadata development projects underway or that have progressed to an early stage of implementation. Indeed, there has been a flood of recent literature on metadata for managing and accessing the electronic resources of specific communities such as education information or government information. There is, for example, a recent special issue of *Government Information Quarterly* on the development of the Government Information Locator System. This issue of *GIQ* 'offers several perspectives on the importance of metadata for resource description and discovery' (Moen, 2001). The diversity of metadata projects has led to concern such as that expressed by Milstead and Feldman (1999), who comment that 'probably the biggest stumbling block in the way of orderly development of metadata is the sheer number of different metadata projects'.

The diversity of metadata projects can be seen in El-Sherbini's survey of a range of ongoing projects (El-Sherbini, 2001). She undertook the survey 'to identify what types of metadata exist and how they are used' and 'to compare and analyse the elements of a selected group of them 'to show how they are related to MARC 21 metadata formats'. El-Sherbini points out that hundreds of metadata projects have been identified, and many of them can be found through the IFLA web page (ww.ifla.org). In her report El-Sherbini describes the following projects (see the appendix to this paper for more information on each):

- the Library of Congress Be-Online Project
  http://lcweb.loc.gov/rr/business/beonline/beohome.html
- the Alexandria Digital Library Project
  www.alexandria.ucsb.edu
  - the following three art projects:
    - Categories for the Description of Works of Art
    - www.getty.edu/research/institute/standards/cdwa/
    - Art, Design, Architecture, and Media Information Gateway (ADAM)
      http://adam.ac.uk/
    - Project Runeberg

>   www.lysator.liu.se/runeberg/
* the Australia New Zealand Land Information Council land and geographic data directory system for Australia and New Zealand www.anzlic.org.au/
* the US Government Information Locator Service (GILS) www.access.gpo.gov/su_docs/gils/
* the Dublin Core Metadata Initiative http://dublincore.org/
* the Federal Geographical Data Committee's *Content standards for digital geospatial data* www.fgdc.gov/metadata/contstan.html
* the Colorado Digitization Project http://coloradodigital.coalliance.org/
* MARC 21 metadata formats www.loc.gov/marc/
* Program for Cooperative Cataloguing Core Level metadata www.loc.gov/catdir/pcc/2001pcc.html.

Resource Description Framework

The diversity of metadata standards is evident from the above list, which covers only a small percentage of metadata projects. As mentioned above, the large number of metadata projects underway has led to considerable concern about the duplication of effort and the lack of common standards, which may lead to problems in establishing interoperability between systems.

The Resource Description Framework (RDF) is an international standard that is currently being developed by the World Wide web Consortium (www.w3.org/RDF/) and other bodies 'to address the problem of incompatible metadata standards' (Pottenger, Callahan and Padgett, 2001). According to Miller (1999), 'RDF may be thought of as superstructure that supports the consistent encoding and exchange of standardized metadata and in so doing assures the interchangeability of separate packages of metadata defined by different resource description communities'. RDF uses XML (eXtensible Markup Language) to provide a common syntax, that is, a common set of rules (similar to rules of grammar) for expressing meaning. Thus it supports the reuse and exchange of different vocabularies across metadata communities. Pottenger, Callahan and Padgett explain that RDF creates a single format for widely varying metadata by

> using the XML namespace at the beginning of a record to give a pointer to a resource that has all the information about the metadata fields used in the record itself. Once a list of references has been given, providing a format for each metadata scheme to be used in the record, the XML tags in the data model are set up to include both the name of the tag, and the metadata model to which this tag belongs. Basically, instead of forcing all records to fit into a common scheme, a Resource Description Framework augments each tag with information about the metadata scheme to which it belongs.
>
> (Pottenger, Callahan and Padgett, 2001)

By imposing a structure on the metadata, RDF should make it easier for organizations to mix and match metadata elements. For example RDF would make it possible for an organization to mix and match Dublin Core and Encoded Archival Description elements that describe the same aspects of an electronic resource, or to reuse metadata elements that emanated from an organization in another metadata community. As Pottenger, Callahan and Padgett (2001) point out, 'this sort of mechanism is pivotal for the creation of federated digital libraries on a large scale'.

Registry programmes

RDF is not the only tool aimed at reducing duplication of effort and aggregating the benefits of the metadata work being done in various communities. Several national and international organizations have undertaken registry programmes that attempt to manage the knowledge developed among metadata projects. Heery and Wagner (2002) explain that 'the motivation for establishing registries arises from domain and standardization communities, and from the knowledge management community'. Duval et al. (2002) note the current importance of research involving metadata registries, which are expected to 'provide the means to identify and refer established schemas and application profiles, potentially including the means for machine mapping among different schemas'. They add that registries will also likely 'contain, or link to, important controlled vocabularies from which the values of metadata fields can be selected' (Duval et al. 2002).

The Schemas Project is an example of a registry research project. The project, which was funded as part of the European Commission's Information Society Technologies Programme, ran from February 2000 until December 2001. The Project was based on the premise that

> no single type of metadata can suit every application, every type of resource, and every community of users. Rather, the broad diversity of potential metadata needs can best be met by a multiplicity of separate but functionally focused metadata packages, or schemas.
>
> (UKOLN, 2000)

The Schema Project's purpose was 'to look at this diverse and often confusing landscape of new and emerging metadata standards from the viewpoint of project or service implementers who must use these standards to design their own interoperable schemas' (UKOLN, 2000). The Schemas Project has published quarterly overviews 'of worldwide progress in the metadata field, which includes work on metadata sets, schemas, frameworks, registries, and the tools needed to create and use all of these' (UKOLN, 2000). The eighth of these reports was published in February 2002, in combination with the Standards Framework Report. This latter report was 'aimed at mapping this wide diversity of metadata standardisation to application areas and to provide information to implementers about what is going on, what they could use and where they can find information about how to use certain standards'. The Schemas registry is available online at www.schemas-forum.org/registry/desire/index.php3

In a recent article in *D-Lib Magazine* Heery and Wagner (2002) explore the roles of metadata registries in relation to the Semantic Web Activity of the World Wide Web Consortium (W3C). The idea behind the Semantic Web is that it is

> an extension of the current Web in which information is given well-defined meaning, better enabling computers and people to work in cooperation. It is the idea of having data on the Web defined and linked in a way that it can be used for more effective discovery, automation, integration, and reuse across various applications. For the Web to reach its full potential, it must evolve into a Semantic web, providing a universally accessible platform that allows data to be shared and processed by automated tools as well as by people.
>
> (World Wide Web Consortium, 2001)

Heery and Wagner discuss a research project involving three prototype registries that were developed by the Dublin Core Metadata Initiative Registry Working Group 'to demonstrate and evaluate application scope, functional requirements, and technology solutions for metadata registries' (Heery and Wagner, 2002). None of the three prototypes resolved all of the problems identified with maintaining a registry involving diverse metadata communities that have developed schemas with differing levels of semantic complexity and serving differing user needs. However, the project resulted in valuable analyses of the associated problems and the difficulties in achieving solutions to those problems.

Application profiles

Application profiles are yet another initiative aimed at maximizing value from the diversity of metadata schemas. Heery and Patel (2000) define application profiles 'as schemas which consist of data elements drawn from one or more namespaces, combined together by implementors, and optimised for a particular local application'. They describe how the work on application profiles provides a context that allows the Dublin Core metadata elements to be combined with other metadata element sets. Duval et al. (2002) explain that application profiles allow the designers of metadata schemas to assemble and combine metadata elements from multiple schemas into a compound schema. The resulting schemas can therefore be tailored to the needs of specific applications and communities, allowing elaboration of locally important metadata elements while maintaining the potential of interoperability with the original base schemas. According to Duval et al. (2002),

> One of the benefits of this approach is that communities of practice are able to focus on standardizing community-specific metadata in ways that can be preserved in the larger metadata architectures of the Web. It will be possible to snap together such community-specific modules to form more complex metadata structures that will conform to the standards of the community while preserving cross-community interoperability.

Another important benefit of application profiles is that they 'provide the means to express the principles of modularity and extensibility' (Duval et al., 2002) in metadata design. Duval et al. explain that modularity makes it possible for designers of metadata schemas to create new groupings of metadata based on existing schemas. It also allows them to take advantage of best practice from the diverse range of metadata communities rather than reinventing elements that may already exist. Duval et al. (2002) also state that extensibility refers to the principle that metadata schemas can be extended to include elements that will accommodate 'local needs or domain-specific needs without unduly compromising the interoperability provided by the base schema'. Thus, while application profiles make it possible for different communities to group metadata elements from other communities with their own metadata (i.e. modularity is supported), they can also add new elements to meet their own needs (i.e. extensibility is supported). Overarching both modularity and extensibility, however, is the need for interoperability between and within communities.

Metadata and interoperability

Interoperability relates to the way systems interact with each other or the way products interact with each other or with systems. According to a definition supplied by Paul Miller (2000b) from the whatis.com website, interoperability refers to 'the ability of a system or product to work with other systems or products without special effort on the part of the customer'. Moen suggests that, when speaking of information systems, 'the ability of two systems and their applications to work together effectively to exchange information in a useful and meaningful manner is a basic definition of interoperability' (Moen, 2001). Miller states that, with regard to electronic information resources, interoperability involves a range of 'flavours', including: technical interoperability, semantic interoperability, political/human interoperability, inter-community interoperability, legal interoperability and international interoperability (Miller, 2000b). Metadata, without doubt, is a component of the first two of these 'flavours' of interoperability.

A metadata schema can be a technical standard in that it can provide the container for the transmission of information about information between systems, allowing for interoperable machine manipulation of that information. A metadata schema can be a semantic standard in that it can prescribe the labels for naming the characteristics of the information being described so that they are applied consistently within and across systems. The other 'flavours' are relevant to the development and application of metadata. For example, the political/human flavour will impact on which metadata schema a community selects as the basis for the description and discovery of its electronic resources, and on how well the metadata is applied. And, the inter-community flavour will impact on the range of partnerships and solutions that are sought across sectors, such as museums, libraries and archives. Bemoaning the tendency for communities to be inward looking and to adapt or create new standards for local use, Gill and Miller (2002) comment:

> Although local or adapted standards are certainly better than no standards, this approach can significantly diminish the value of a digital cultural collection by limiting its interoperability with the wider networked world.

The use of formal or recognized standards that have been developed with input from a broad community will lead to far greater sharing of resources than will the use of local or adapted standards. In October 2001 the Dublin Core Metadata Element Set was accepted as a formal standard, ANSI/NISO Z39.85, by the National Information Standards Organization (NISO) in the USA. NISO is an accredited non-profit association that 'identifies, develops, maintains, and publishes technical standards to manage information' (NISO, 2001). Formal standards will make it more readily possible for organizations to contribute to and benefit from distributed digital collections or to share their electronic resources through other management structures. Libraries have known this for a long time, given that the original MARC format was developed in the 1960s. The introduction of the MARC standard allowed libraries far greater sharing of their resources through organizations such as OCLC, which simplified the exchange of machine-readable bibliographic records and the use of those records in the creation of computerized union catalogues.

Metadata and preservation

The preservation community is cognizant of the value of metadata for preservation management purposes with regard to digital information resources. The OCLC/RLG Preservation Working Group, for example, states that preservation metadata 'is the information necessary to carry out, document and evaluate the processes that support the long-term retention and accessibility of digital content' (OCLC/RLG Working Group on Preservation Metadata, 2002). In a state-of-the-art review of preservation metadata the Working Group stated:

> Effective management of all but the crudest forms of digital preservation is likely to be facilitated by the creation, maintenance, and evolution of detailed metadata in support of the preservation process. For example, metadata could document the technical processes associated with preservation, specify rights management information, and establish the authenticity of digital content. It can record the chain of custody for a digital object, and uniquely identify it both internally and externally in relation to the archive in which it resides. In short, the creation and deployment of *preservation metadata* is likely to be a key component of most digital preservation strategies.
>
> (OCLC/RLG Working Group on Preservation Metadata, 2001b)

The Working Group refers to the two functional objectives of preservation metadata. The first objective is to provide preservation managers with sufficient information to maintain the digital object's bit-stream over the long term, while the second objective is to ensure that the content of a preserved object can be 'rendered and interpreted, in spite of changes to access technologies'. Each of these functional objectives has been addressed in a specific recommendation (see OCLC/RLG Working Group on Preservation Metadata, 2001a and OCLC/RLG Working Group on Preservation Metadata, 2001b).

Another notable preservation metadata project was the UK-based Cedars (CURL Exemplars in Digital Archives) Project which ran from mid-1998 through March 2002 (Day, 2002). The Cedars Project, part of the Electronic Libraries Programme, was a collaborative effort involving the universities of Leeds, Oxford and Cambridge (Cedars

Project, 2002). A useful tool from the project is the *Cedars guide to preservation metadata*, which is aimed at providing 'a brief introduction to current preservation metadata developments and introduc[ing] the outline metadata specification produced by the Cedars project' (Cedars Project, 2002).

There are still other uses to which metadata can be put, such as capturing information relevant for managing libraries of digital objects, e.g. technical metadata to identify the accuracy of the digital object's reflection of the original, or to ensure that the data for the digital object can be refreshed periodically and migrated when necessary (Library of Congress, n.d.). Metadata can also be used for documenting decisions pertaining to the selection of resources for inclusion in, or exclusion from, a digital library collection. However, these types of metadata reflect local concerns more than community concerns, and they are less likely to generate or follow formal standards. Nonetheless, their application in a given project may still be worth investigating.

Conclusion

This chapter has attempted to capture the words from a digital spinning top to enable us to understand the current issues and recent developments pertaining to metadata. However, the metaphor of the digital spinning top suggests that the words are spinning in a blur. The fact that the spinning top is *digital* means that the words on it are changing rapidly, so the task of keeping up to date is far more difficult.

One of the great difficulties with metadata is that it can be both simple and complex at the same time. The Dublin Core, for example, was created from a minimalist perspective with a view to allowing the metadata to be determined by the authors of electronic resources rather than by information professionals. While the Dublin Core identifies 15 elements as the basis for metadata records aimed at resource description and discovery, it does not prescribe the source or format for the content of those elements. For example, the Dublin Core does not specify where the name of the creator should be obtained, nor does it specify whether it should be entered as 'first name, last name' or 'last name, first name'. To standardize the decision-making process in establishing the metadata for books the library community uses the second edition of the *Anglo-American cataloguing rules*, which by any stretch of the imagination is not a simple set of rules to understand and apply.

When an organization becomes involved with metadata creation, regardless of the schema that it selects, it must make decisions about which staff members will add the metadata to their resources, and what rules they will use to standardize their decisions. The development of policies during the initial stages of a digitization programme is critical to successful decision making. However, as mentioned in the introduction, this chapter is not intended to be a guide on how to create metadata or how to establish a metadata project. For that type of guidance, works such as Hudgins, Agnew and Brown's *Getting mileage out of metadata: applications for the library* (1999) and *Guidelines for the creation of content for resource discovery metadata* from the National Library of Australia and State Library of Tasmania should be consulted.

These are exciting times. Libraries, archives, museums and galleries are digitizing resources to make them more accessible to their users. From these processes new ways of making those resources accessible are being found, and new users are finding and exploiting the riches of our cultural and heritage institutions. For example, the Electronic Text Center at the University of Virginia Library received 'about 16 million web "hits" on 5 million documents from over 1 million visitors – predominantly K12 and general public users' for access to online texts and downloadable books in November 2001 (University of Virginia, 2002). In 2000–2001 the Electronic Text Center provided electronic books and related material to users from 190 different countries.

In Australia the Picture Australia site (www.pictureaustralia.org)

provides free access to over 565,000 digital images of Australian people, places and events. These digital images are stored in a distributed digital library which links together the digital collections of 16 libraries, museums, galleries, archives, universities and other cultural agencies in Australia and abroad, and they can all be searched at the same time. The most exciting aspect of Picture Australia and other digitization projects available freely on the web is that they can be searched by primary and secondary school children, by university students and 'serious' researchers, by the curious and by people who are simply surfing the web and stumble across resources that inform them about their own or other cultures. These people can be anywhere in the world, searching at any time of the day. And metadata is one of the vital keys to managing these resources so that they can be discovered and accessed.

References

Association for Library Collections and Technical Services. Task Force on Metadata (2000) *Final Report, June 2000,* available at www.ala.org/alcts/organization/ccs/ccda/tf-meta6.html. (accessed 23 May 2002)

Association of American Publishers (2000) *Metadata standards for ebooks*, Washington, DC: Association of American Publishers.

Bakewell, D. et al. (1999) *An infrastructure for forestry metadata creation and access*, Ottawa: Natural Resources Canada, available at www.pfc.forestry.ca/eosd/resources/Background/doug_bakewell_metadata_infrast.html (accessed 5 June 2002)

Brand, A. (2001) CrossRef Turns One, *D-Lib Magazine*, **7** (5) , available at www.dlib.org/dlib/may01/brand/05brand.html. (accessed 25 May 2002)

Brisson, R. (1999) The world discovers cataloging: a conceptual introduction to digital libraries, metadata and the implications for library administrators, *Journal of Internet Cataloging*, **1** (4), 3–30.

Caplan, P. L. (1995) You call it corn, we call it syntax-independent metadata for document-like objects, *The Public Access Computer Systems Review*, **6** (4), 19–23.

Cedars Project (2002) *Cedars guide to preservation metadata*, Leeds: University of Leeds.

Consultative Committee for Space Data Systems (2001) *Reference model for an open archival information system (OAIS): draft recommendation for space data systems standards: CCSDS 650.0-R-2*, available at www.ccsds.org/documents/pdf/CCSDS-650.0-R-2.pdf (accessed 2 June 2002)

Cousins, T. R. and Dominick, W. D. (1978) The management of data bases of data bases. In *The information age in perspective: proceedings of the ASIS annual meeting*, vol. 15, White Plains, NY: Knowledge Industry Publications for American Society for Information Science, 75–8.

Dack, D. (2001) *Persistent identification systems: report on a consultancy conducted by Diana Dack for the National Library of Australia: May 2001*, available at www.nla.gov.au/initiatives/persistence/PIcontents.html (accessed 2 June 2002)

Davidson, L. A. and Douglas, K. (1998) Promise and problems for scholarly publishing, *Journal of Electronic Publishing*, **4** (2), available at www.press.umich.edu/jep/04-02/davidson.html (accessed 25 May 2002)

Day, M. (2001) *Metadata in a nutshell*, London: United Kingdom Office for Library and Information Networking, available at www.ukoln.ac.uk/metadata/publications/nutshell/ (accessed 25 May 2002)

Day, M. (2002) *Metadata: CURL Exemplars in Digital Archives (Cedars)*, London: United Kingdon Office for Library and Information Networking, available at www.ukoln.ac.uk/metadata/cedars/

(accessed 2 June 2002)

Dublin Core Metadata Initiative (2000) *History of the Dublin Core Metadata Initiative*, available at
http://dublincore.org/about/history/
(accessed 26 May 2002)

Duval, E. et al. (2002) Metadata principles and practicalities, *D-Lib Magazine*, **8** (4), , available at
www.dlib.org/dlib/april02/weibel/04weibel.html
(accessed 2 June 2002)

Education Network Australia (2000) *Metadata homepage*
http://standards.edna.edu.au/metadata/
(accessed 25 May 2002)

El-Sherbini, M. (2001) Metadata and the future of cataloging, *Library Review*, **50** (1), 16–27.

Fietzer, W. (1999) Technical services and the sociology of metadata, *Technicalities*, **19** (6), 1, 13–15.

Gill, T. and Miller, P. (2002) Re-inventing the wheel? Standards, interoperability and digital cultural content, *D-Lib Magazine*, **8** (1), available at
www.dlib.org/dlib/january02/gill/01gill.html
(accessed 5 June 2002)

Gilliland-Swetland, A. J. (2000) Setting the stage. In Gilliland-Swetland, A. J. (ed.) *Introduction to metadata: pathways to digital information*, 2nd edn, Los Angeles, CA: Getty Research Institute, available at
www.getty.edu/research/institute/standards/intrometadata/
2_articles/index.html
(accessed 2 June 2002)

Gorman, M. (1999) Metadata or cataloging? A false choice, *Journal of Internet Cataloging*, **2** (1), 5–22.

Gradmann, S. (1998) *Cataloguing vs. metadata: old wine in new bottles?* Paper presented at the 64th IFLA General Conference (Amsterdam), available at
www.ifla.org/IV/ifla64/007-126e.htm

(accessed 6 June 2002)

Hall, B. (2001) Executive Summary. In *Getting up to speed on e-learning standards*, Brandon-Hall.com, available at www.brandon-hall.com/htstandards.html (accessed 5 June 2002)

Heery, R. and Patel, M. (2000) Application profiles: mixing and matching metadata schemas, *Ariadne*, **25**, available at www.ariadne.ac.uk/issue25/app-profiles/ (accessed 2 June 2002)

Heery, R. and Wagner, H. (2002) A metadata registry for the Semantic Web, *D-Lib Magazine*, **8** (5), available at www.dlib.org/dlib/may02/wagner/05wagner.html (accessed 26 May 2002)

Hillman, D. (2001) *Using Dublin Core*, available at http://dublincore.org/documents/usageguide/ (accessed 26 May 2002)

Hudgins, J., Agnew, G. and Brown, E. (1999) *Getting mileage out of metadata*, Chicago: American Library Association.

Lange, H. R. and Winkler, B. J. (1997) Taming the internet: a work in progress, *Advances in Librarianship*, **21**, 47–72.

Larsgaard, M. L. (1996) Cataloging planetospatial data in digital form: old wine, new bottles — new wine, old bottles. In Gilliland-Swetland, A. J. (ed.), *Geographic information systems and libraries: patrons, maps, and spatial information: papers presented at the 1995 Clinic on Library Applications in Data Processing*, Urbana-Champaign, IL: University of Illinois at Urbana-Champaign, Graduate School of Library and Information Science, 17–30.

Library of Congress. Network Development and MARC Standards Office. *METS: Metadata Encoding and Transmission Standard official web site* www.loc.gov/standards/mets/ (accessed 3 June 2002)

Lipkin, J. A. (1982) Designing data resources directories – real world considerations, information interaction, In *Information*

Interaction: *proceedings of the 45th ASIS Annual Meeting*, Vol. 19, Columbus, Ohio, October 17–21, White Plains, NY: Knowledge Industry Publications.

Liston, D. M. and Dolby, J. L. (1982) Metadata systems for integrated access to numeric data files, *Drexel Library Quarterly*, **18** (3/4), 147–60.

Mainelli, T. (2002) New view of passport data leaves some customers unhappy with what they see, PCWorld.com, available at www.pcworld.com/news/article/0,aid,100084,tk,wb052002x,00.asp (accessed 2 June 2002)

Medeiros, N. (1999) Driving with eyes closed: the perils of traditional catalogs and cataloging in the internet age, *Library Computing*, **18** (4), 300–5.

Miller, E. (1999) Making progress: the Resource Description Framework (RDF), *Journal of Internet Cataloging*, **1** (4), 53–8.

Miller, P. (2000a) I am a name and a number, *Ariadne*, **24**, available at www.ariadne.ac.uk/issue24/metadata/ (accessed 24 May 2002)

Miller, P. (2000b) Interoperability: what is it and why should I want it?, *Ariadne*, **24**, available at www.ariadne.ac.uk/issue24/interoperability/ (accessed 24 May 2002)

Milstead, J. and Feldman, S. (1999) Metadata: cataloging by any other name, *Online* (January/February), 32–8, 40.

Moen, W. E. (2001) The metadata approach to accessing government information, *Government Information Quarterly*, **18**, 155–65.

National Information Standards Organization (2001) *About NISO*, available at www.niso.org/about/ (accessed 2 June 2002)

National Library of Australia *Picture Australia* www.pictureaustralia.org/ (accessed 3 June 2002)

National Library of Australia and State Library of Tasmania (n.d.)
Guidelines for the creation of content for resource discovery metadata,
available at
www.nla.gov.au/meta/metaguide.html
(accessed 2 June 2002)

OCLC/RLG Working Group on Preservation Metadata (2001a) *A
recommendation for content information: October 2001*, available at
www.oclc.org/research/pmwg/contentinformation.pdf
(accessed 3 June 2002)

OCLC/RLG Working Group on Preservation Metadata (2001b)
*Preservation Metadata for digital objects: a review of the state of the art:
January 31, 2001*, available at
www.oclc.org/research/pmwg/presmeta_wp.pdf
(accessed 3 June 2002)

OCLC/RLG Working Group on Preservation Metadata (2002)
OCLC/RLG Preservation Metadata Working Group, available at
www.oclc.org/research/pmwg/
(accessed 2 June 2002)

Olson, N. (1997) *Cataloging internet resources: a manual and practical
guide*, 2nd edn, available at
www.oclc.org/oclc/man/9256cat/toc.htm
(accessed 24 May 2002)

Paskin, N. (n.d.) *A message from the Director of the International DOI
Foundation (IDF)*, available at
www.doi.org/welcome.html
(accessed 25 May 2002)

Pottenger, W. M., Callahan, M. R. and Padgett, M.A. (2001)
Distributed information management. In Williams, M. E. (ed.)
Annual review of information science and technology, vol. 35,
Medford, NJ: Information Today for the American Society for
Information Science, 79–113.

Potter, L. and Bide, M. (1999) *User passports and visas: understanding
the role of identification metadata*, London: Book Industry
Communication on behalf of the British National Bibliography

Research Fund.

Publishers International Linking Association (2002) *CrossRef: the central source for reference linking*, available at
www.crossref.org
(accessed 4 June 2002)

Taylor, A. (1999) Where does AACR2 fall short for internet resources?, *Journal of Internet Cataloging*, **2** (2), 43–50.

UK Office for Library and Information Networking (2000) *Schemas: project objectives*, available at
www.schemas-forum.org/project-info/objectives.htm
(accessed 26 May 2002)

University of Virginia. Library. Electronic Text Center (2002) *Etext quick facts 2001–2002*, available at
http://etext.virginia.edu/stats/
(accessed 3 June 2002)

Vellucci, S. L. (1998) Metadata. In Williams, M. E. (ed.) *Annual review of information science and technology*, vol. 33, Medford, NJ: Information Today for the American Society for Information Science, 187–222.

Vernon, R. D. and Rieger, O. Y. (2002) *Digital asset management: an introduction to key issues*, Ithaca, NY: Cornell University, Office of Information Technologies, available at
www.cit.cornell.edu/oit/Arch-Init/digassetmgmt.html
(accessed 19 May 2002)

Widerkehr, R. R. (1982) Methodology for representing data element tracings and transformations in a numeric data system, *Drexel Library Quarterly*, **18** (3/4), 161–76.

World Wide Web Consortium (2001) *Semantic Web Activity statement*, available at
www.w3.org/2001/sw/Activity
(accessed 26 May 2002)

Appendix:
Descriptions of Metadata Projects Analysed by El-Sherbini (2001)

The following list provides a brief description of the metadata projects analysed by El-Sherbini, along with a URL that can be accessed for more information about each project. Each URL was last accessed on 2 June 2002.

- The Library of Congress Be-Online project
  (http://lcweb.loc.gov/rr/business/beonline/beohome.html)
  Now called Be-Online+, it 'concentrates on business and economics-related material' (El-Sherbini, 2001, p.17). This project uses the Library of Congress Program for Cooperative Cataloguing (PCC) Core Record to catalogue websites and provide authority-controlled access points to them in records that are made up of required MARC fields but are not based on AACR2.
- The Alexandria Digital Library Project
  (www.alexandria.ucsb.edu)
  Focuses on spatial data, especially 'remote-sensing imagery such as aerial photographs and satellite images' (p.18). This project started out using MARC 21, AACR2 and Library of Congress Subject Headings 'but has added to or departed from these as appropriate for specific material types'.
- The following three art projects:
  - Categories for the Description of Works of Art: is aimed at art and architecture specialists and provides a structure for describing works and electronic images of them using categories rather than rules as the basis of its standards.
    www.getty.edu/research/institute/standards/cdwa/
  - Art, Design, Architecture, and Media Information Gateway

(ADAM): uses established standards including AACR2 and Olson's *Guide* (see Olson, 1997), and the Art and Architecture Thesaurus.

http://adam.ac.uk/

— Project Runeberg: a digitization project for making Nordic texts and information about those texts available on the web.

www.lysator.liu.se/runeberg/

• The Australia New Zealand Land Information Council land and geographic data directory system: a joint project for Australia and New Zealand which has developed its own metadata elements using a framework that is consistent with the US Federal Geographic Data Committee's guidelines on Digital Geospatial Metadata and with the Australia New Zealand Standard on Spatial Data Transfer AS/NZS 4270.

www.anzlic.org.au/

• The Government Information Locator Service (GILS): a tool for identifying US government information resources. This service has a complex metadata structure highly influenced by the MARC and Z39.50 communities.

www.access.gpo.gov/su_docs/gils/

• The Dublin Core Metadata Initiative: a simple resource description record of 15 core elements for describing electronic resources.

http://dublincore.org/

• The Federal Geographical Data Committee's *Content standards for digital geospatial data*: a complex metadata system for helping researchers locate data sets for use in geographical information systems.

www.fgdc.gov/metadata/contstan.html

• The Colorado Digitization Project: a collaboration involving libraries, archives, historical societies and museums in providing access to digitized historical and cultural resources about Colorado. This project uses seven elements from the Dublin Core, plus other metadata elements that are useful to specific institutions for describing resources.

http://coloradodigital.coalliance.org/
- MARC 21 metadata formats: international standards 'for the representation and communication of bibliographic and related information in machine readable form' (p.20). These representations are created using established standards such as AACR2, LCSH and the Dewey Decimal Classification. The framework for MARC is complex, involving record structure, content designators and the actual content.

    www.loc.gov/marc/
- Program for Cooperative Cataloguing Core Level metadata: an alternative to MARC that establishes a mid-way standard between minimal and full catalogue records (p.20). This standard establishes a core level for creating 'useful records' but provides some flexibility for cataloguers to add extra elements when deemed necessary.

    www.loc.gov/catdir/pcc/2001pcc.html

15

Beyond today's search engines

Christopher Brown-Syed

Introduction

It has often been said that the internet, and the world wide web, are chaotic and require organization to be useful. The information studies community, which for our purposes can be taken to include librarians, documentalists and archivists, has responded to this perceived problem by attempting to catalogue, classify and otherwize organize the internet. With a few notable exceptions, the computing community has taken a different approach, attempting to develop computer programs and to exploit new techniques in machine intelligence to create comprehensive indexes of the internet and the web.

Search engines are familiar and widely used tools. An informal poll of incoming Master of Library Science students in any institution in North America, Europe or the Pacific Rim will probably reveal that more university graduates are aware of having used internet search engines than of having used traditional library catalogues. However, anyone who has spent time researching specific topics using search engines can attest to their strengths and weaknesses, and to the frustration that occasionally occurs when one feels certain that needed information exists, but can be located only by wading through scores or even thousands or hundreds of thousands of irrelevant responses to a simple query. Internet search engines have always excelled at delivering high recall – vast numbers of responses to given queries. Recently, they have become much better at delivering precision – smaller retrieval sets, whose top-ranked elements make better sense in the context intended by the questioner. As well, because of the popularity of the web, existing engines have had to curtail the frequency of visits to sites and the depth of their delving at individual sites. There are simply too many servers and too many documents on the web to index each and every page effectively.

While most users have favourite search engines, the need to improve retrieval for the web is apparent to all. Several strategies present themselves; this chapter examines each in turn. They include several possible sorts of enhancements to the retrieval process, such as:

- improving the efficiency and effectiveness of search engines intended for traditional text-oriented web searches
- adding to display options by giving users the options to arrange hit lists chronologically or topically, or to expand or limit searches
- enhancing existing engines so that they mirror the bibliographic research process more closely
- creating repositories that contain authoritative information by emulating traditional bibliographic databases
- involving humans in the process as peer reviewers or as research assistants, thereby providing more services and features of a traditional print library
- applying citation analysis and artificial intelligence techniques to search engine design.

Additionally, web designers can improve site content to expedite its eventual retrieval by improving the content of web pages using metadata and by taking advantage of design techniques intended to improve access for hearing- or sight-impaired users, which may have spin-off value for indexing, such as the use of alternate text captions for pictures.

Since these points about design are not the focus of this chapter, it will suffice to observe in passing that it takes only a few minutes to add metadata to a document and alternate text tags for its pictures, and the resulting accessibility and indexing pay-off could be considerable. A caption on a picture renders that graphic element just as suitable for indexing as text. For the promise of metadata to be fulfilled, web-composing tools which prompt for meta-tags, or which generate suggested tags by examining document content, must be made available, and search engines must be set up to take advantage of the tags.

History of the internet

Before briefly discussing each retrieval option we must take a short historical detour. In the computing industry the concept of legacy software and hardware is well understood. The term applies to ageing tools which an organization has retained because it is too expensive or disruptive to replace them. Arguably, 'legacy thinking' also exists, and to understand how this might be, we must examine some notions of web organization that apply more fittingly to its precursors – the file systems of individual computers, and pioneering internet services such as Gopher and the Wide Area Information Service (WAIS).

Because the internet has been in existence for only a few decades, it is sometimes easy to overlook its history. This is especially so because one tends to view the internet as cutting-edge technology. Work began in the late 1960s on the communications protocols which eventually came to define its inner workings. An early specification of the Arpanet protocols, RFC 5, was issued in 1969. The very first RFC (Request for Comments document) was also issued that year. The Transmission Control Protocol and Internet Protocol, which now govern basic network communications underlying services such as the web, were summarized in RFC 793, edited by Jon Postel, in 1981. Its most popular utility, the world wide web, has been available for a mere ten years. Talk of a history of the internet will seem premature to those who have been part of its development and use over the few short decades which span the text-only world described in Hiltz and Turoff's classic, *The Network Nation* (1978), which described the largely non-commercial realms of Arpanet, Bitnet and Usenet, and the contemporary world of dot coms and online videos.

Today's internet developed piecemeal and is the result of dozens of sometimes serendipitous developments. It is perhaps due to the efforts of volunteers like Postel that any primary source record of its history exists. The primary sources for a technical history of the internet are easily accessible, thanks largely to the work of such chroniclers. Postel's dedication is evident in his careful editing of many of the hundreds of RFC documents that make up the de facto body of standards used by internet programmers.

Legacies of the virtual directory model

The task of anticipating what lies beyond web search engines is daunting, because the internet continues to evolve, as computing costs continue to decrease, and while computing power and transmission bandwidth become cheaper. Today's search engines, however, exist in response to retrieval problems perceived, in some cases, almost two decades ago. As early as 1992 librarians were looking toward the internet as a possible virtual or distributed library. The University of Minnesota's Gopher, along with complementary or competing schemes such as WAIS, figured highly in university-campus-wide information system plans.

A Gopher implementation produced a screen display at the user's client machine which resembled a file folder (directory) display. When a user searched a Gopher server, that server returned a display which looked very much like a file folder or directory display on the user's workstation. This display was referred to as a 'virtual directory listing'.

> While documents (and services) reside on many servers, Gopher client
> software presents users with a hierarchy of items and directories much

> like a file system. The Gopher interface is designed to resemble a file system since a file system is a good model for organizing documents and services; the user sees what amounts to one big networked information system containing primarily document items, directory items, and search items (the latter allowing searches for documents across subsets of the information base). (Anklesaria et al., 1993)

Local file systems are hierarchical, commencing with a root directory and consisting of numerous subfolders. On Unix machines the root is the directory with no name, called '/'. On Microsoft systems each disk drive has its own root directory, identified by the drive letter, for example, 'C:\'. Disk file systems are inherently organized and hierarchical, and the assumption of Gopher's designers was that both users and designers would adhere willingly to this type of imposed structure. It seemed logical that designers would want to keep related collections of documents in separate dedicated file folders and that a virtual directory of the internet could be set up which allowed local autonomy and took advantage of client server software, but which preserved this inherently hierarchical and highly organized arrangement.

Tools like Gopher and WAIS held out great promise during the early 1990s, as is evident in this prediction made personally to this author in 1992 by Australian systems librarian Tony Barry: 'Increasingly, we will be looking at the net globally, which increasingly is looking like gopherspace with a few odd library things hanging off it. [I]ndividuals are increasingly publishing direct to the net, and tools to control this in an automated fashion are developing rapidly.'

Campus Wide Information Systems (CWIS) were designed with this model in mind. Users would enter the site via a main menu, and proceed through series of menus of increasingly fine resolution until they located the desired documents. A good parallel would be a telephone answering system, which offers users numeric choices of options: for example, 'to find a nearby bank branch, press 2. To open a new account, press 3'. Librarians, to whom this model should be strikingly familiar because it resembles classification tables, may cling to the notion of its continued applicability in the web era and beyond. To do so would be unsatisfactory. While the virtual directory model underlies the web, the inclusion of hyperlinks within documents, and the introduction of search engines which index individual pages and disregard the directory structures as organizing principles, allow users to bypass the hierarchical structures imposed by file systems on web servers.

It is perhaps useful to note, in passing, that some network services have altered almost beyond recognition. For example, an early web application, DejaNews, was originally intended to expedite searches of the emphatically non-commercial Usenet newsgroups. It subsequently grew into a web portal with a highly commercial flavour. With Google's acquisition of Usenet group content, the Usenet postings of a decade ago may be found listed among web hits. This is perhaps not surprising, since newsgroups have been accessible from browser screens for many years.

Usenet itself was always organized by topic. Early guides to netiquette by volunteers like Eugene Spafford and Brad Templeton cautioned Usenet posters against advertising, spamming, and similar offences. Highly organized Usenet groups were voted into existence through a process described in Conner-Sax and Krol (1999). Usenet content was and is divided into areas of interest – recreation, science, society, and so forth. Commercial messages, mainly for trading in second-hand goods, were to be posted only in the group 'misc.forsale'. However, network abuses such as spam and rampant cross-posting of messages could only be deterred by peer pressure, and in Usenet's heyday the number of users was relatively small, so the technique worked well. Notorious abusers were reported to their systems administrators, who chided them or even stopped their accounts. With millions of internet users, this practice now seems as archaic as the frontier justice of the American West.

Additionally, the web presents new indexing problems simply because it is not an English-language text-only body of data. Usenet

News and Gopher grew out of the 80-column ASCII character set and ISO Roman alphabet of the world before Windows. When improved graphics and ample bandwidth for their transmission became possible, web designers rushed to add photographs, graphics, sound clips and eventually WebCams and streaming video content. However, search engines were designed to index text. The need for multilingual support, the ability to use other character sets, to provide translation services, and to make the web accessible to those with hearing or visual impairments, have presented additional challenges for designers in recent years.

The web is not a database

It is tempting for librarians to view the web in the same light as they do subscription databases, such as Ovid, Dialog, ERIC or CARL, or a MARC-encoded catalogue. This view is misleading for several reasons. One is that there is no 'record structure' in a web page. This is unlike a database, which has separately searchable fields for authors, titles, subjects, and so on.

The beauty – and perhaps the shortcoming – of Hypertext Mark-up Language (HTML), which remains the basis of web pages, lies in the fact that the codes and the text are intermixed. As a result, web pages are platform independent – they can be displayed on almost any computer, from a palmtop to a desktop Macintosh or a Microsoft Windows machine, to a large Unix machine. Since an HTML file (web page) is divided only into a basic head section, normally invisible to the user, and a body section containing the content to be displayed, one way to improve retrieval would be to add tags to the head. The Dublin Core Metadata Initiative does precisely this – providing ways to identify authors, keywords and abstracts for documents. Sadly, neither web content designers nor search engine builders seem fully aware of the potential of metadata. Newer schemes, such as eXtensible Mark-up Language (XML), offer ways of dividing pages into identifiable sections. However, by definition each designing organization can invent its own sections, with the result that consistence is even harder to achieve.

The web is inherently wild

While the internet, which is becoming synonymous with the web, is often described as 'the world's library', the two have little in common. A library is inherently organized by human and machine endeavour. Its contents have been carefully selected, classified, catalogued and arranged for access. By contrast, the internet is more like a billion stacks of books, journals, photos and half-finished essays lying on the floors of untidy readers' dens, scattered about the world. Moreover, the comparison is unfair.

Unlike 'virtual directories' such as Gopher, the web was never meant to be hierarchical and organized. Observing that there were things the human brain did, like forming ad hoc associations, which computers had heretofore been unable to mimic successfully, Tim Berners-Lee set out to create an inherently disorganized document production and distribution system which would take advantage of a handful of simple protocols and which would offer the ability to connect any object on any web server to any other object, anywhere on the internet.

> The dream behind the Web is of a common information space in which we communicate by sharing information. Its universality is essential: the fact that a hypertext link can point to anything, be it personal, local or global, be it draft or highly polished. There was a second part of the dream, too, dependent on the Web being so generally used that it became a realistic mirror (or in fact the primary embodiment) of the ways in which we work and play and socialize. That was that once the state of our interactions was on line, we could then use computers to help us analyze it, make sense of what we are doing, where we individually fit in, and how we can better work together.
>
> (Berners-Lee, 1998)

The advantage of Berners-Lee's design is that any word, picture, sound clip, video or any digitized object can be related to any other such object, and links among objects can be determined using a search engine, and the locations of related objects assembled in real time. Since objects are updated regularly by their authors or creators, there is no need to construct and maintain Gopher-style menus. The contents of the web change from minute to minute as designers around the world update documents or photos.

Constructing a 'virtual directory listing' of the entire web using human, or even automated, means is doomed from the start to immediate obsolescence. However, search engines attempt to do just that by taking snapshots of the contents of various servers at pre-determined frequencies. Search engine robots or spiders or crawlers are programs which roam their way about the internet, indexing selected pages from known server sites.

Beginning with the server root directories, spiders copy some or all of the data in the web pages (HTML documents) they find there, following some or all of the links embedded in these pages, and stopping at some arbitrary point. Sub-folders on the sites may also be explored. Each spider follows its own rules regarding the number of documents or the number of links it will follow at any given site. Thus, when AltaVista's spider and Google's spider visit a particular site, each may index a slightly different set of pages.

The web includes real-time components

More recently, the widespread use of interactive methods, such as Active Server Pages (ASP), Java and Perl scripts, has made web objects even more transient and ephemeral. Using ASP or Java, a web designer can arrange for a unique page to be created for the user in real time. As a result, two people accessing the same site, but submitting different queries to that site, or submitting requests from geographically disparate locations, may see different versions of the site's pages, created for them on the fly.

Because software on the server side can detect the user's time and space characteristics by looking up that user's Internet Protocol (IP) number in a directory, or identify the user's preferences based on prior access patterns, a user in Canada who begins by accessing the Google search engine at google.com may be directed automatically to google.ca. A user in the UK may be directed to the bookstore site, amazon.co.uk, and presented with suggestions for reading based on previous orders, while a user from Australia may access the same site and be shown an entirely different screen. It is therefore impossible to provide a decisive snapshot of the internet – it is continually creating itself, and in real time.

Cataloguing the web – directory projects

Likewise, directories of the web constructed by humans with the aid of machines can only hope to create snapshots. In the best traditions of classical librarianship, printed guides such as Krol's *Whole internet* (Krol, 1992), Hann and Stout's *Internet yellow pages* (Hann and Stout, 1994) and the Gale Group's Cyberhound series have sought to catalogue the web. Pioneering online projects like Project Gutenberg have sought to provide quality content by digitizing printed works, while the groundbreaking Internet Public Library (http://ipl.sils.umich.edu) and its successors, such as the Librarian's Index to the Internet (www.lii.org) attempt to keep up with the daunting task of cataloguing broader web content. Of these directory projects, Yahoo! occupies pride of place.

Constructing and maintaining directories is labour intensive but offers the advantage of something much like peer review or collection management techniques in the scholarly and library worlds. Humans can select and include links only to sources that they consider authoritative or of particular interest, creating interactive documents that serve the same purpose as do printed pathfinders to libraries or archival finding aids. The subject expertise of the cataloguer is paramount. However, new selection criteria are necessary.

The rules that govern the selection of print and professionally produced audiovisual materials have been recognized and codified in procedure manuals like the *Anglo-American cataloguing rules*. The title pages of books, and the reputations of journal or book publishers, citation analysis, as well as many other hallmarks of authority and authenticity, exist in the print and broadcast media realms. Almost by definition, additional tools are needed for identifying reliable web documents. With the exception of the case of online offerings of previously printed articles and books, no similar hallmarks exist for the web. This is because, as Berners-Lee hoped, the web contains documents of all sorts and all levels of completion.

Walker and Janes, whose *Online retrieval* is used widely in library schools, have included an observation about the shortcomings of pre-coordinate indexing in the second edition of their work (Walker and Janes, 1999). With web content growing at such a pace, just keeping up with outdated links could prove too labour intensive for the builders of tomorrow's directories without the aid of link-checking programs.

With the advent of the web the need for online guides became more and more apparent. Yahoo!, described as the 'de facto catalogue' of the internet by Walker and Janes (1999), is the most celebrated of these guides. Yahoo! began as an online version of Krol or Hann, a list of links selected and categorized by humans. However, automated tools such as AltaVista soon emerged. These classic search engines aimed at recall rather than precision, to use terms common in information retrieval circles. That is, the aim of the original versions of engines like AltaVista was to index most of the internet and to allow users to retrieve as many hits as possible, given the terms they selected. With its inception in 1995, AltaVista earned pride of place among classic search engines. It was one of the first services to establish regional repositories in an attempt to provide more relevant results.

> Since the early days, other notable inventions to the engine include the first-ever multi-lingual searches on the Internet. We are also proud of

Babel Fish — the Web's first Internet translation service that translates words, phrases, and entire Web sites online in Spanish, French, German, Portuguese and Italian. More recently, we launched Photo Finder, an image search technology. Other improvements include phrase detection, spell check, and natural language capabilities.

(AltaVista, 2002)

The model of a classic search engine involves a fact-finding spider program, an indexing program and a retrieval engine. By examining the web server usage logs at any particular site, systems administrators can determine the frequency with which that site is visited by spiders. On average, a modest academic website can expect about 100 visits from spiders over the course of a month or so. The frequencies of spider visits vary with the indexing company, Google, AltaVista, and the others. Not every page on a site is indexed. Some engines, for example, may examine only the files in the server root directory, not in subdirectories. Others may record the site's home page, and those pages directly linked to it, while others may follow links on subsequent pages as well.

Website managers may also request that spiders not index certain pages, or even entire subdirectories. The Robots Exclusion Protocol allows designers to list names of folders or files in a file called 'robots.txt' situated in the server root directory – the folder containing the site's home page, which is generally called 'index.html'. The metadata tag <meta name='robots' content='noindex'> can be included in the header of any web page, and constitutes a request that a specific page not be indexed regardless of its location, and the 'nofollow' parameter can request that robots not follow links on the page. So far, the major search engines appear to be obeying robots exclusion requests. As a consequence, robots cannot and do not claim to index the entire web.

Because new pages on a site will not be indexed immediately, there will be delays before they appear in search engines. If the order in which hits are presented is unclear to users, much old material may appear first on a retrieval screen. More recently, engines such as Teoma, now a product of Ask Jeeves, attempt to categorize and cluster retrieved hits and to rank them according to their evident popularity as links on similar sites. This process is much like citation or co-citation analysis, and it presents additional complications. Since new pages will not have been cited, their apparent authority among peer sites will be indeterminate. Thus, a preponderance of older material will appear on search sets.

According to Teoma's website (www.teoma.com), the technique was devized in 2000 by a team from Rutgers University. After performing a standard search, Teoma uses three additional techniques: 'Teoma organizes sites into naturally occurring communities that are about the subject of each search query. These communities are presented under the heading "Refine" on the Teoma.com results page. This tool allows a user to narrow or sharpen their search (AskJeeves, 2002). In other words the engine attempts to establish a subject heading for a site. The software then uses a technique called 'subject-specific popularity' to establish a co-citation pattern among related sites and to rank hits by popularity. Additionally, the system provides a list of other authoritative sites associated with the topic.

Another approach to improving the quality and authoritative results of web searches, including improving precision, involves the dovetailing of humanly constructed directories with search engines. Humanly created directories, while at best partial, continue to make sense as adjuncts to search engines. An example of this approach is the Open Directory Project, which dubs itself 'The Definitive Catalog of the Web'.

> The Open Directory follows in the footsteps of some of the most important editor/contributor projects of the 20th century. Just as the Oxford English Dictionary became the definitive word on words through the efforts of [. . .] volunteers, the Open Directory follows in its footsteps to become the definitive catalog of the Web. The Open Directory is the most widely distributed data base of Web content classified by humans.

Its editorial standards body of net-citizens provide the collective brain behind resource discovery on the Web. The Open Directory powers the core directory services for the Web's largest and most popular search engines and portals, including Netscape Search, AOL Search, Google, Lycos, HotBot, DirectHit, and hundreds of others.

(www.dmoz.org)

While it is true that search engines are incorporating human catalogues into their regimens, the reverse is equally true. Although Yahoo! began as a humanly constructed directory, it has long incorporated search engines into its database-building mechanisms. One example is Inktomi, based in Foster City, California. Inktomi's web server lists a constellation of important clients: 'Inktomi's customer and strategic partner base includes such leading companies as America Online, AT&T, Compaq, Dell, Hewlett-Packard, Merrill Lynch, Microsoft, Nokia, Sun Microsystems and Yahoo!' (www.inktomi.com). Inktomi's spiders regularly traverse the web, and the data they glean may end up on various sites. In other words, when using a particular directory or search engine on the web, one may be reaping the harvest of a myriad of unmentioned spiders.

While search engines may provide immediate answers to specific known-item queries, users conducting extensive research require additional facilities. Search engines are only as good as their content, the frequencies with which they are updated, their abilities to purge or revise outdated information, and their abilities to deliver context-sensitive and highly relevant material.

Meta-engines, such as HotBot, Dogpile and others, which perform simultaneous searches against the databases of individual engines, offer the advantage of making less work for the searcher. They attempt to trade off recall and precision by offering a sample of results from many different search engines, but limiting the number of hits from any given search engine. Thus the results Dogpile or Hotbot present as having come from, say, Excite or Lycos, may differ from those obtained from individual searches of those services. As well, services like Google offer a few screens of hits with high prospects of relevance, and allow users to access additional hits if desired. The implication in both cases is that researchers seeking high recall might be best advised to conduct searches individually.

Specialized content repositories

In still another approach, some online search vendors are turning to the classic model of bibliographic indexing and abstracting sources such as Dialog Ovid, or ERIC, by creating pre-selected sets of high-quality documents. Some of these offer subscriptions, like their pre-web relatives, while others offer pay-per-view solutions. For example, Northern Light is now almost totally given over to pay per view, and its content is derived increasingly from printed journalistic and academic sources. Viewers are shown 'summaries', which may not be full abstracts or articles, and invited to purchase the source. In other words Northern Light is Dialog for US$2.50 a reprint.

Meanwhile, Google has adopted a novel approach to the problem of research assistance. Users are invited to pay US$4.00 and up for the answers to their research questions. This marketing gimmick is new, and time will prove its success or failure.

It should be mentioned that many of the world's newswires and newspapers, along with professional societies and scholarly print journals, are now making reprints available at a cost to the user. As the library increasingly becomes less distinguishable from the web, the web, or parts of it, appears to be coming closer to the classical model of the printed book, paper or journal. This trend is evident if one visits the CNN Time-Warner websites – users are shown teasers for video clips, but must subscribe in order to view them.

Artificial intelligence

The past few decades have also seen advances in artificial intelligence. As early as the 1980s, programmers were experimenting with logic-based languages such as Prolog and Lisp to create expert systems. An expert system mimics the actions of a human by following established rules and recognizing members of lists of known items. For example, a Prolog-like program might consist of rules like: 'mammal:- warm-blooded.', and accompanying lists of known mammals: 'mammal:- [cow, dog, horse]'. As the program encountered more animals filling the condition of warm-bloodedness, it would add them to the list. Newer artificial intelligence techniques involve wilder ideas, such as neural networks. At the turn of the 21st century, neural network techniques are being applied in the analysis of web documents, citation patterns and more. A good description of neural network computing may be found on the web pages of McMaster University's Department of Psychology (http://claret.psychology.mcmaster.ca/NeuralComp/index.html):

> The brain is comprised of billions of simple processing units (nerve cells) operating in parallel, in a massively interconnected network. Artificial neural network models, or ANNs, have the ability to emulate this remarkable power to perform complex parallel computations, and what's more, they have the ability to learn.
>
> ANNs and closely related machine learning techniques have been applied successfully to a wide range of signal processing and classification problems, such as automatic handwriting recognition for postal-code scanning, fraud detection in credit card use, real-time speech recognition for consumer electronics, DNA and protein sequencing.

There is at least one product on the web market today which uses the notion of thousands of tiny units operating independently to achieve the results of a larger engine.

> Colony is a system of distributed autonomous Agents and applications, designed to manage one of the biggest problems facing corporations, organizations, and individuals today – getting maximum advantage from the enormous, rapidly expanding and enriched networked

> resources of content (data, information and knowledge) stored in many formats across all the platforms in their organization and beyond. (Colony, 2002)

The metaphor of an ant colony is a powerful one. It comes with the attendant promise of exploiting self-organizing phenomena and the emergent properties of systems – of bringing order from chaos. The expectation raised here is that each tiny software agent, aware of only its immediate purpose, could be sent out to retrieve a piece of information from the internet, much like an ant might be given the mission of bringing back a piece of leaf. Back at the colony, sense is made of thousands of bits of leaves – they become dinner or construction material.

In real life an ant sent out to find leaves might retrieve a tasty drop of honey if it chanced upon it. For the metaphor to be fully realized as a software solution, this 'target of opportunity' behaviour would be of no consequence, for the main colony software would presumably sort out such matters. If we have thousands of ant-like workers at our disposal, odds are good that most will perform their assigned tasks. Those who do not will either be innovators with a knack for finding something better, or misguided ones whose efforts we can ignore.

Information from the web is increasingly touted as being 'mission critical' to organizations. With advances in artificial intelligence techniques, and the commercial impetus to engage them, we can expect many more artificial-intelligence-enhanced search tools in the near future.

The impact of convergence

Bibliographic databases, highly organized and quality assured, arose during the era of the mainframe computer. The early internet was conceived in the distributed processing environment which coincided roughly with the introduction of minicomputers. The web reached prominence only with the spread of end-user computing and became useful when internet service providers could rely on increased bandwidth, processing speed and inexpensive mass storage. Cable modems and ISDN lines now allow millions of network users to access the internet swiftly at home. What will be the impact upon librarianship when that power and speed become universally portable – when the Personal Digital Assistant (PDA) includes both server and client software?

Tomorrow's PDA promises to combine the functions of the mobile telephone, cellular videophone, e-book, still and movie camera, personal organizer, and the e-mail and web browser. Indeed, the first generation of products like these are already appearing on the market in 2002, offered by mobile phone vendors like Nokia.

The prospect of millions of users creating web and e-mail content, operating streaming video servers on trams and street corners, and hoping to retrieve documents 'on the go', is staggering. We have yet to meet the challenge of constructing a 'world library', even given the relatively stationary and limited number of internet servers in existence so far. When millions of users around the world carry the equivalent of a web server and client in a palm-sized packet, the task of organizing this ubiquitous resource will grow exponentially.

The best hope for maximum recall and optimum precision appears to rest with a combination of increasingly intelligent and dialogue-capable search engines, and with services that mimic the activities of traditional librarians and researchers. These extended functions may be addressed by integrating artificially intelligent agents into search engines, or by making human librarians as easily accessible as telephone directory assistance operators. The extent to which the automated can meet the demands both of ready reference seekers and of users requiring in-depth research remains a matter for conjecture.

There is a final reason to express hope for the PDA solution – it promises to bring developing nations a reasonable hope of achieving information equity. One initiative aimed at just this is the Simputer. Designed to run under Linux, and to exploit speech recognition and synthesis for those who cannot read, this initiative from Bangalore aims at putting the internet into the hands of farmers and villagers in developing nations (Simputer Trust, 2000). When palmtop solutions like the Simputer are coupled with advances in human and artificial web searching, Tim Berners-Lee's dream of a ubiquitous mechanism for the exchange of knowledge and the formation of social understanding may indeed fall within our grasp.

Conclusion

This chapter began with a discussion of the notion of a distributed file system and the inapplicability of hierarchical organization to the inherently chaotic world of the web. Various suggestions for better organization, pre-coordinate or post-coordinate, have been mentioned. While it may be desirable for people to collect and categorize information on the web, constructing pathfinders, directories and enhanced databases, this path presents both advantages and drawbacks. On the plus side, it preserves the notions of controlled vocabulary and peer review. On the minus side, it inevitably means that valuable resources will be excluded either because they are new or because they are overlooked.

Metadata offers a partial answer, but while meta-fields like keywords can be mandated, even machine generated at document creation time, there is no agreed vocabulary that could ensure standardization across languages and disciplines. Moreover, for metadata to succeed both authors and search engine designers must cooperate in exploiting it.

Better retrieval techniques, including improved search engines, recourse to human experts, and the application of artificial intelligence may prove the better course. This approach offers the advantages of fewer complications for end-users but presents its own challenges. Traditional search engines have provided great recall, but like humans, artificial search engines may miss good resources, while presenting the illusion of comprehensiveness. As well, they have been less than satisfactory at providing precision.

Better engines – ant colonies, neural nets, faster computers, and regional distributed web indexes all hold great promise. Spelling correction, 'see from' and 'see also' references, and facilities to refine or expand searches are gaining popularity. Multilingual support, adequate translations and speech synthesis and recognition present challenges for the immediate future that lies beyond today's search engines.

References

AltaVista (2002) *About AltaVista: company background*, available at www.altavista.com/sites/about (accessed 1 May 2002)

Anklesaria, F. et al. (1993) *University of Minnesota. Network Working Group, Request for Comments 1436. The internet Gopher Protocol (a distributed document search and retrieval protocol)*, (March), available at http://freesoft.org/CIE/RFC/Orig/rfc1436.txt (accessed 1 May 2002)

AskJeeves (2002) *Teoma Search: search with authority – the Teoma difference*, available at www.teoma.com (accessed 1 May 2002)

Berners-Lee, T. (1998) *The world wide web: a very short personal history*, available at www.w3.org/People/Berners-Lee/ShortHistory.html (accessed 1 May 2002)

Colony (2002) *Welcome to TryColony.com: the intelligent management of knowledge assets*, available at www.trycolony.com (accessed 1 May 2002)

Conner-Sax, K. and Krol, E. (1999) *The whole internet: the next generation. a completely new edition of the first and best user's guide to the internet*, Cambridge, MA: O'Reilly and Associates.

Hann, H. and Stout, R. (1994) *The internet yellow pages*, Berkeley, CA: Osborne McGraw-Hill.

Hiltz, S. R. and Turoff, M. (1978) *The network nation: human communication via computer*, Reading, MA: Addison-Wesley Publishing Company [revised edition 1993].

Inktomi Corporation
www.inktomi.com
(accessed 1 May 2002)

The Internet Public Library
http://ipl.sils.umich.edu
(accessed 1 May 2002)

Koster M. (1996) *A method for web robots control*, Network Working Group, Internet-Draft, 4 December, available at www.robotstxt.org/wc/norobots-rfc.html
(accessed 1 May 2002)

Krol, E. (1992) *The whole internet user's guide and catalog*, Sebastopol, CA: O'Reilly and Associates.

Librarian's Index to the Internet
www.lii.org
(accessed 1 May 2002)

McMaster University. Department of Psychology (2001) *Neural Computation*, available at http://claret.psychology.mcmaster.ca/NeuralComp/index.html
(accessed 1 May 2002)

Open Directory Project
www.dmoz.org
(accessed 1 May 2002)

Simputer Trust
www.simputer.org
(accessed 1 May 2002)

Walker, G. and Janes, J. (1999) *Online retrieval; a dialog of theory and practice*, 2nd edn, Englewood, CO: Libraries Unlimited.

16

Are we information providers or the information police? The uneasy marriage between access and security

Shadrack Katuu

Introduction

The mantra for information professionals, regardless of their disciplinary affiliation (librarians, content managers, knowledge architects, museum curators, archivists), is that we provide education for information use and access to information. This is evident in many professional discussions, in journals and the objectives of our professional bodies. A cursory glance at the mission, vision, aims, or objectives of some of the associations reveals that the theme of education and access is recurring. Table 16.1 presents excerpts of information gleaned from the associations' byelaws, constitutions and general information resources.

Table 16.1 *The access imperative for information professionals*

Association	Vision, mission, mandate, aims or objectives
American Library Association	'Provide leadership for the development, promotion, and improvement of library and information services and the profession of librarianship in order to enhance learning and ensure access to information for all' (ALA, 2002).

Table 16.1 (continued)

American Society for Information Science and Technology	'Advancing knowledge about information, its creation, properties, and use; . . . increasing public awareness of the information sciences and technologies and their benefits to society' (ASIS&T, 2001).
ARMA International	'To enable them [information professionals] to use their skills and experience to leverage the value of records, information, and knowledge as corporate assets and as contributors to organizational success' (ARMA, 2002).
International Council on Archives	'Facilitating the use of archival documents by making their contents more widely known and encouraging greater ease of access' (ICA, 2002).
International Council on Museums	'Professional cooperation and exchange dissemination of knowledge and raising public awareness of museums' (ICOM, 2002).
International Federation of Library Associations and Institutions	'Promote high standards of provision and delivery of library and information services' (IFLA, 2001).
Society of American Archivists	'To foster a better public understanding of the nature and value of archival operations and holdings' (Society of American Archivists, 1997).
Special Library Association	'And to promote and improve the communication, dissemination, and use of such information and knowledge for the benefit of libraries or other educational organizations' (SLA, 2000).

But times are changing. Beyond the traditional resources that have been available in libraries and other information organizations, information professionals are now grappling with information generated electronically. Such information is made available through access to digital collections, remote databases, online catalogues and intranet portals. In addition traditional services have radically changed, allowing for the provision of e-services, which may include electronic reference, user education or information literacy, and e-request/e-renewal/e-recall services. All of these focal points affect access to information.

Writing in 1998 on access controls in library environments, Morgan expressed a sense of confusion and despondency with the challenges of meeting professional obligations. He stated:

> I'm confused. I've spent the majority of my career as a librarian trying to provide free and equal access to information, and now I'm being told to implement access control systems, security measures, filtering software, and authorization/authentication gateways. Are we providing information services for our mutual benefit, or are we becoming the information police?

Information provision – access

In this new networked information age 'we are seeing the emergence of a web of inter-organizational trust relationships in support of . . . information access, implemented and expedited through new authentication and access management systems' (Lynch, 1999). According to the Internet Society (2002), access to information on networks is seen as a multifaceted public policy issue. It may not be easy to distinguish the various facets, since in most cases they tend to overlap. However, for purposes of this discussion, it is possible to identify three main facets: intellectual access, social-economic access, physical access.

Intellectual access

Intellectual access involves empowering the end-user to comprehend the information received. With the increasing availability of technology users have become techno-literate. However, as one librarian has noted, 'computer fluency is not equal to information literacy' (Barron, 2002). While computer literacy requires skills that enable an individual to use computers, software applications, databases, and other technologies to achieve a wide variety of work-related or personal goals, information literacy requires a broader set of abilities. For an individual to be information literate, one has to be able to 'recognize when information is needed and have the ability to locate, evaluate, and use effectively the needed information' (ALA, 1989). In the contemporary environment, characterized by rapid technological developments and the proliferation of information resources, information literacy empowers users to access and use resources more effectively. It is also important to bear in mind that individuals with different physical and mental capacities will need different degrees of assistance in order to attain a desired standard of literacy.

Social access

The social-economic aspect of access has been characterized by discussion of the digital divide. Bridges.org (2001) defines the digital divide as 'the divide that exists between countries and between different groups of people within countries, where there is a wide division between those who have real access to information and communications technology and are using it effectively, and those who don't'. This concept encompasses the issue of ensuring equity based on initiatives that may include the change of legal and regulatory frameworks, taking account of socio-cultural factors (such as race and gender), improving both macro- and micro-economic environments, as well as advocating the political willingness of national and regional governing bodies.

Physical access

However, the main interest of this chapter is physical access, which entails ensuring that the infrastructure is both available and physically accessible. This infrastructure includes computer hardware and software, telecommunications infrastructure, network hardware and software. In its simplest form, physical access entails users visiting information centres and having the right to use the resources. In a simpler world physical access would entail opening hours, signage, fees and the availability of the physical resources. However, in the digital world the challenges of physical access become more complicated. A user in an academic library, for instance, may want to access electronic information in the institution's own network. That is simple enough if all the hardware and software are in place to provide this information. The scenario becomes more complicated when the user requires information that is available in another institution's network. The home library will need to have negotiated access to the other institution in advance. In instances where this information is at a cost, then that will have to be settled before access is provided. In other circumstances the user is not even within the premises – and indeed the remote user is becoming increasingly common, and a major concern where access is concerned.

In all such situations there is a web of activities required in order to allow users to access the network. The growing complexity of access requirements is accompanied by an increased number of access activities. Illustrating these is difficult without first delineating the components of an access system. Van Halm (1999) argues that there are four major components: the access interface system, access control system, pricing system and billing/clearing/charging system.

- *Access interface system*. This consists of one or more databases that provide links to various resources and provides access to internal and external information resources. This system may be a library home page or even a web portal, which is closely monitored and regularly updated for enriched access.
- *Access control system*. This controls or restricts access to contents using local user data as well as session controls. This may consist of on/off campus access, session controls, authentication, authorization and user accounting.
- *Pricing system*. This permits the configuration of models for priced content. Each model differs depending on the user community, the kind of content used and the databases accessed.

- *Billing and clearing system.* This allows the charging of individual users for services and clears the accumulated amounts between user communities. It may provide such features as shopping carts, different means of payments, invoice configurations, and the billing of different services by means of a uniform interface.

Often there are two main groups of users, the unlicensed and the licensed. Unlicensed users may be able to access the free services on the networks, such as searches. Most often they will interact with the first two components of the access system. However, licensed users who have access to most of the institutional and remote resources will interact with all the components of the access system.

Computer workstation and network security

The reality of networking, especially where there are large numbers of users, is the high degree of vulnerability to all kinds of premeditated as well as inadvertent misuse. US government computers, for example, 'that handle trillions of dollars in tax refunds and Social Security benefits remain vulnerable to cyber-attacks' (CNN, 2002). A recent report by one US government agency found various access weaknesses in networks, among them:

- inadequate access controls, such as passwords and locks
- poorly administered system software, including duplicate or obsolete programs
- poor segregation of employee duties, giving certain employees more control than they should have had
- no comprehensive security program covering the entire system
                                                                    (CNN, 2002).

Is it any wonder, then, that reports abound on how often network security has been compromised in government and private firms? Some of the more prominent security breaches in recent years include:

- the CIA
- the FBI
- the Japanese government Science and Technology Agency, who had census data erased in January 2000 (Jacobsen, 2000)
- Yahoo, Amazon, eBay, CNN.com, Buy.com, ZDNet, E*Trade, and Excite.com, who suffered denial-of-service attacks in February 2000 that disabled their services for a brief period (Neumann, 2000a)
- the Swedish National Board of Health and Welfare, whose web server was attacked (Lindqvist, 2000).

In this kind of environment it is common to find that information integrity and security can be severely compromised in both a networked and a computer standalone situation. A recent report on system and network survivability classifies the types of computer misuse techniques that afflict network security (Neumann, 2000b) In this report Neumann argues that of these different classes, the most prevalent seem to be bypasses and pest programs.

Bypasses

Neumann states that 'bypassing may involve circumvention of existing controls, modification of those controls, or improper acquisition of otherwise denied authority, presumably with the intent to subsequently misuse the acquired access rights' (Neumann, 2000b). The bypassed controls may be operating systems authentication mechanisms, database control systems or even firewalls. The cases of unauthorized access are often the result of system and usage flaws such as:

- inadequate identification, authentication and authorization of users, tasks or systems
- improper initialization, being the result of improper initial domain selection or security parameters, improper partitioning, or other embedded operating system parameters in application memory space

- improper finalization resulting from, among other activities, incompletely handled aborts, accessible vestiges of de-allocated resources, incomplete external device disconnections
- inadequate encapsulation where there has been lack of information hiding, accessibility of internal data structures, alterable audit trails, mid-process control transfers, as well as hidden or undocumented side effects (Neumann, 2000b).

Most malicious bypasses may be motivated by boredom, the lure of a challenge, greed – for data or information that can be gleaned from it – specific anger at the library or organization or individual, or generalized anger at the whole world (Ives, 1996).

Pest programs

The story of the Kitsap Regional Library network (www.krl.org/) in Washington state illustrates the extent of havoc that pest programs can create. A pest program, NIMDA worm, struck the system early one morning, mutilated more than 5000 web pages, paralyzed the web servers, infected all other file servers, and rendered several computers, the online public access system and web mail systems unusable. The system librarians states that 'every member of the computer room staff had spent full time on this [project] for 7 days . . . . The cost to the library in eradicating this virus, including actual time and deferred projects, plus inconvenience to staff and public alike, exceeded $10,000' (Schuyler, 2002).

Safeguards

According to David Ives, implementing simple and inexpensive security procedures can reduce security risks in information professionals' environments by 75 to 95%. In his view there are three key points that all staff charged with security should remember.

- Think like the enemy – examine your system with an eye towards answering two questions: What damage could I do if I wanted to? What are the weaknesses of my system?
- Defence in depth. Two or more methods used jointly are more effective than one.
- Dynamic defence – always assume that computer security procedures can be improved and that they need to be improved continuously (Ives, 1996).

Ives further adds that it is important to ensure that the organization's security programme and procedures are *not*

- negotiable – they must apply to all users equally and in the strongest possible manner
- consensus based – there are certain absolutes to security, and this reality cannot be changed by any committee-based agreements
- open to compromise – no special events, employees, or circumstance can ever justify compromising a security programme
- decentralized – a central security authority, for example, the computer support group, must hold the sole authority and responsibility for implementation, enhancement and enforcement of an organization's security programme and procedures (Ives, 1996).

In order to protect information on desktop and networked environments, there are numerous protection mechanisms that could be employed. Some of the safeguards that information professionals have suggested include:

- Updating virus software. From his experience at the regional network in Washington state with the pest program NIMDA, Michael Schuyler (2002) suggests ensuring that anti-virus software is updated very frequently.
- Using firewalls. There is an argument that 'security threats are more likely to come from inside our libraries than to come from

outside via the Internet' (Charnetski, 1998). A potential solution for this is to use firewalls, which should be implemented according to a 'design based on the organization's particular vulnerabilities and security requirements' (Charnetski, 1998).

- Authenticating access. Authentication and access control measures might include user name and password authentication, cookie file authentication, smart cards, digital certificates, cryptography and encryption. Some of these matters are discussed by Kim Guenther (2001).

- Using security software. At a single workstation level the dangers include rearranging workstation desktops, tampering with files, downloading files to the hard drive, launching program files and booting from the A: drive (Schneider, 1999). One solution to this is using security software such as WinU from Bardon Data Systems (www.bardon.com/winu.htm), Fortres (Fortres Grand Corporation) (www.fortres.com/) or WinSelect (Hyper Technologies) (www.winselect.com/pages/products/WS50_Info.htm?B122=More+Info) to hide all the drives that the public do not need and disable anything that cannot be hidden (Schneider, 1999). When the computer is launched, these programs allow one to hide all drives and the network neighbourhood apart from the A: drive, removing the operating systems desktop and leaving a simple menu with one or two options on it. The programs also ensure that the interface for all computers remains simple and consistent with the same colours, configurations (e.g. home page and print settings), programs and files for every user.

Challenges: maintaining privacy and free speech

However, any discussion about safeguarding information collections is inevitably also a discussion about safeguarding the privacy of users, because most security activities compromise this privacy to some extent. A study released by the Privacy Foundation reveals that an average of one of every three of the 40 million employees in the USA using e-mail or the internet is monitored (Shankland, 2001).

Here are some of the ways in which the desire to protect can compromise privacy:

- Network administrators could easily install 'sniffing' equipment whose main duty may be to monitor data traffic and ensure that the network operates at a maximum. However, these same tools could be used to trace packets of data, reveal their contents and hence compromise an individual's right to the exclusivity of their work. There is also a high risk in this scenario for information to be tampered with.
- The process of researching users' information-seeking behaviour using various data collection techniques may cause concern about how and what kind of information is obtained from the users. The analysis of transaction logs, for example, may be one way of collecting comprehensive representation of user activities. However, if these logs can be associated with individual users, then it may also be possible to learn about individual surfing habits. Do information professionals have this right?

In a case that took centre stage in December 2001, a British subject tried to blow up American Airlines Flight 63 from Paris to Miami. Subsequent reports stated that 'reconstructing the virtual and physical trail that Reid left in the . . . neighborhoods of Paris, Brussels and Amsterdam could be key to understanding and prosecuting the networks suspected of assisting him' (Rotella, 2002). Prosecutors state that the 'investigation has centered largely on emails . . . exchanges with suspected confederates during long hours hunched over computer terminals in cyber cafes in Brussels and Paris' (Rotella, 2002). Detectives have raided these cafés and gathered evidence from computer hard disks (BBC, 2002). Though the suspect spent a lot of time in cyber cafés, which are not necessarily considered traditional information centres in a Western sense, these e-mails could as well have been sent from a public library. The question is, how would public librarians have responded to this?

There are several technologies central to the debate about the need to balance the government's use of technology to track criminals with the public's right to privacy, as Table 16.2 indicates.

Table 16.2   *Electronic 'Eavesdropping' Devices (Olsen and Hansen, 2001)*

Device	Use	Latest information
Carnivore (now renamed DCS1000)	The FBI in the USA has been using the technology, which allows it to tap communications that go through internet service providers.	In July 2001 the US House of Representatives passed a bill requiring federal law enforcement officials to release details about the use of electronic surveillance systems. The bill now goes to the Senate for approval.
Echelon	This is a surveillance network shared by US, British, Canadian, Australian and New Zealand intelligence agencies. It allegedly can capture communications from a variety of sources, including satellite and undersea cable lines.	The European Parliament released a report in May 2001 confirming the existence of Echelon, saying it has been capable of intercepting messages since 1978.
Fingerprint scanning	The biometrics technology prints and transmits fingerprints electronically to identify people.	The Alabama Bureau of Investigation has said that it would use the technology to submit fingerprint and demographic records electronically during background checks.
Key-logger system	The technology captures the keystrokes made on a computer and can be used to discover encrypted passwords.	In *United States v. Nicodemo S. Scarfo* a federal judge ordered prosecutors to show him documents describing how the system works. In closed session, the judge will decide whether the government is protected under the Classified Information Procedures Act, which prevents disclosure for reasons of national security (Olsen, 2001).

Clarke (1999) argues that there are at least four dimensions of privacy: privacy of the person, privacy of personal behaviour, privacy of personal communication, and privacy of personal data. This discussion has mainly highlighted issues related to information privacy, which Clarke sees as combining two dimensions, communications privacy and data privacy. He defines information privacy as 'the interest an individual has in controlling, or at least significantly influencing, the handling of data about themselves' (Clarke, 1999). The process of protecting information privacy is a complicated one of finding appropriate balances between privacy and multiple competing interests. These interests may belong to a group of people, an organization or society as a whole.

Related to information privacy is the need to safeguard intellectual freedom. The American Library Association's Office for Intellectual Freedom (2002) defines this as 'the right of every individual to both seek and receive information from all points of view without restriction. It provides for free access to all expressions of ideas through which any and all sides of a question, cause or movement may be explored.' A computer science professor who is a researcher in cryptography maintains that the current encryption regulations in the USA violate his free speech and prevent him and colleagues from conducting legitimate research (Bowman, 2002) – and similar views are undoubtedly held in many countries.

In this kind of environment it is clear that information professionals must maintain a precarious balance between maintaining the freedoms that are expected in their respective countries, including the right to privacy, while also investing in security technologies to protect their considerable investment in information resources and information technology. 'Over time, a calmer attitude will prevail and tighter security will be in place where it needs to be' (Well, 2002).

Conclusion

Regardless of the contentious issues surrounding the provision and policing of information, the risks involved in providing access in an increasingly networked environment are both real and potentially crippling if there are no safeguards for security. 'Along the continuum of total security to open access lies a viable path, but getting to that path is not easy' (Rezmeierski and Soules, 2000). And one should add that not getting to that path is foolhardy at best and could be catastrophic at worst.

Information policing is a necessary evil if information provision is to continue in a highly networked environment. Often privacy advocates argue that 'there is no guarantee that any security program or set of procedures will address or resolve all potential problems' (Ives, 1996). However, real events in the past demonstrate that not establishing these programmes and procedures could cripple an organization. Though these programmes may not be 100 per cent effective, it is hoped that they 'will be robust enough to discourage attempts at security breaking and will be sufficiently effective to persuade troublemakers to look elsewhere for easier prey' (Ives, 1996).

References

American Library Association (1989) *Presidential Committee on Information Literacy. Final report*, Chicago: American Library Association, available at
www.ala.org/acrl/nili/ilit1st.html
(accessed 23 February 2002)
American Library Association (2002) *ALA Mission, membership, organization – an overview, January 2002*, available at
www.ala.org/alaorg/mission.html
(accessed 23 February 2002)
American Library Association. Office for Intellectual Freedom (2002) *Intellectual freedom and censorship Q & A*, 10 January, available at
www.ala.org/alaorg/oif/intellectualfreedomandcensorship.html
(accessed 1 March 2002)

American Society for Information Science and Technology (2001)
   ASIS&T 2001, available at
   www.asis.org/AboutASIS/asis-mission.html
   (accessed 23 February 2002)
ARMA International (2002) *About ARMA International*, available at
   www.arma.org/about/about.cfm
   (accessed 23 February 2002)
Bardon Data Systems (2002) *WinU*, available at
   www.bardon.com/winu.htm
   (accessed 1 March 2002)
Barron, B. B. (2002) Distant and Distributed Learners Are Two
   Sides of the Same Coin, *Computers in Libraries*, **22** (1), 24–8.
BBC News (2002) *Show bomb suspect 'Sent Suicide Email'*, 19 January,
   available at
   http://news.bbc.co.uk/hi/english/world/americas/newsid_
   1769000/1769927.stm
   (accessed 5 February 2002)
Bowman, L. M. (2002) *Taking on Uncle Sam over encryption*, available
   at
   http://news.com.com/2008-1082-804901.html
   (accessed 1 February 2002)
Bridges.org (2001) *What is the digital divide?* , available at
   http://www.bridges.org/digitaldivide/index.html
   (accessed 23 February 2002)
Charnetski, J. R. (1998) Avoid disaster: use firewalls for inter-
   intranet security, *Computers in* Libraries, **18** (9), 44–7.
Clarke, R. (1999) *Introduction to dataveillance and information privacy,
   and definitions of terms*, available at
   www.anu.edu.au/people/Roger.Clarke/DV/Intro.html#Priv
   (accessed 1 March 2002)
CNN (2002) *GAO: key government computers remain vulnerable*, available at
   www.cnn.com/2002/TECH/internet/02/05/security.government.
   reut/index.html

(accessed 11 February 2002)

Fortres Grand Corporation (2001) *Fortres*, available at
www.fortres.com/products/index.htm/
(accessed 1 March 2002)

Guenther, K. (2001) Knock, knock, who's there? Authenticating
users, *Computers in Libraries,* **21** (3), 54–6.

Hyper Technologies (2002) *WinSelect*, available at
www.winselect.com/pages/products/WS50_Info.htm?B122=
More+Info
(accessed 1 March 2002)

International Council on Archives (2002) *The mission of ICA*, available at
www.ica.org/eng/mission.html
(accessed 23 February 2002)

International Council on Museums (2002) *What is ICOM?* , available
at
www.icom.org/organization.html
(accessed 23 February 2002)

International Federation of Library Associations and Institutions
(2001) *More about IFLA*, available at
www.ifla.org/III/intro00.htm
(accessed 23 February 2002)

Internet Society (2002) *Internet connectivity access*, 8 January, available
at
www.isoc.org/isoc/access/
(accessed 1 May 2002)

Ives, D. J. (1996) Security management strategies for protecting your
library's network, *Computers in Libraries*, **16** (2), 36–41.

Jacobsen, O. J. (2000) *Japanese government websites hacked*, available at
http://catless.ncl.ac.uk/Risks/20.77.html#subj2
(accessed 25 February 2002)

Kitsap Regional Library
www.krl.org/
(accessed 1 March 2002)

Lindqvist, U. (2000) *Swedish 16-year-old arrested 3 hours after web attack*, 5 April, available at http://catless.ncl.ac.uk/Risks/20.87.html#subj4 (accessed 25 February 2002)

Lynch, C. (1999) Authentication and trust in a networked world, *Educom Review*, **34** (4), 60.

Morgan, E. L. (1998) Access control in libraries, *Computers in Libraries*, **18** (3), 38–40.

Neumann, P. (2000a) Distributed denial-of-service attacks, 14 February, available at http://catless.ncl.ac.uk/Risks/20.79.html#subj1 (accessed 25 February 2002)

Neumann, P. (2000b) *Practical architectures for survivable systems and networks (Phase Two final report)*, Menlo Park, CA: Computer Science Laboratory, SRI International, available at www.csl.sri.com/neumann/survivability.pdf (accessed 6 February 2002)

Olsen, S. (2001) U.S. keeps PC surveillance under wraps, 24 August, available at http://news.com.com/2100-1023-272131.html?legacy=cnet (accessed 24 February 2002)

Olsen, S. and Hansen, E. (2001) Terrorist threat shifts priorities in online rights debate, 17 September, available at http://news.com.com/2102-1023-272972.html (accessed 24 February 2002)

Rezmeierski, V. and Soules, A. (2000) Security anonymity, *Educause Review*, **35** (2), 22–9.

Rotella, S. (2002) 'Shoebomber' suspect linked to 2 other plots, *Los Angeles Times*, (3 February), available at www.latimes.com/news/printedition/front/la-000008650feb03.story?coll=la-headlines-frontpage (accessed 5 February 2002)

Schneider, K. G. (1999) Safe from prying eyes: protecting library systems, *American Libraries*, **30** (1), 8.

Schuyler, M. (2002) A serious look at systems security, *Computers in Libraries*, **22** (1), 36–9.

Shankland, S. (2001) Study: web, e-mail monitoring spreads, 8 July, available at http://news.com.com/2100-1001-269584.html?legacy=cnet (accessed 24 February 2002)

Society of American Archivists (1997) *Section I: constitution & bylaws of the Society of American Archivists*, available at http://207.21.198.172/governance/handbook/section1.asp (accessed 23 February 2002)

Special Library Association (2000) *SLA bylaws*, available at www.sla.org/content/SLA/Bylaws/slabylaw.cfm (accessed 23 February 2002)

Van Halm, J. (1999) The digital library as access management facilitator, *Information Services and Use*, **19** (4), 299–305.

Well, N. (2002) What are the hot trends in technology?, 12 February, available at www.cnn.com/2002/TECH/industry/02/12/hot.trends.tech.idg/index.html (accessed 16 February 2002)

Index

Page numbers in *italics* indicate references to figures

Buttenfield, B. 265, 268
Buy.com 367
bypasses, of security 367–8

cable modems 357
Callahan, M.R. 308, 321
Calvino, I. 98
Cambridge University (UK) 327
Campus Wide Information Systems (CWIS) 344–5
Canada
    collaboration in international digitization programmes 297–9
    Digital Cultural Content Initiative 71, 288
        funding for digital programmes 71
        internet access 137, 138
Caplan, P.L. 309
Carnegie, A. 69
Carnivore 372
carpel-tunnel syndrome 38
Carter, D.S. 194, 269
Cast, M. 188, 198
cataloguing
    cataloguing data contrasted with metadata 311
    cataloguing the web with directories 349–54
    connotations of "catalogue" 309
    online catalogues 140–1, 160
    sharing of catalogues 4
    of web resources 163–4
    see also metadata
Categories for Descriptions of Works of Art 319, 337
CAVAL Ltd 225
CD-ROMs 7–9, 18, 30–1, 142
Cedars guide to preservation metadata 328
Cedars Project 327–8
Central Queensland University (Australia) 221
Centre for Research in Library and Information Management, Manchester Metropolitan University 61, 64
Centre for Studies in Advanced Learning Technology, Lancaster University 61–2
CERLIM, Manchester Metropolitan University 61, 64
Cervantes Virtual Library, University of Alicante 72
change, libraries as agents of 244–5
Charnetski, J.R. 369–70
chat 139, 198
Chelton, M. 99
Chen, H. 73
Chen, S. 141
children see younger users
CHILIAS Biblioteca Virtual 148–9
China
    access to computers 137
    internet use 242–3
Chronicle of Higher Education 211
CIA 367
citation analysis 352
citation indexes 274, 275
Citizens' Gateway, Essex Libraries 290–1

Clarke, R. 373
classification, of web resources 162–3, 164
Cleveland Public Library 246
CNN Time-Warner 354
CNN.com 367
co-branding 147
co-citation analysis 352
co-operative programmes 4, 20–1, 246, 248–50, 297–9
    see also consortia
Cochrane, P. 48
Coffman, S. 189, 198
Cohen, D. 216
COINE Project 64
Collaborative Digital Reference Service (US) 143, 249
collaborative programmes 4, 20–1, 246, 248–50, 297–9
    see also consortia
collection development 174, 286
collection management, of digital collections 261–77
Colony 355–6
Colorado Digitization Project 320, 338
Commonwealth Writers Prize 40
community resources, libraries as 63–4
Compaq 353
computer skills 172–3
computer workstation security 366–73
computers, disposal of 38
Computers in libraries 147
conferencing technology 198–9
Congo 83
Conner-Sax, K. 345
conservation 60
consortia 19–20, 75, 83, 196
    see also collaborative programmes
consumer portals 166–7
consumers, and service delivery 214–15
contact centre software 198
content evaluation 263, 270
    see also digital content creation programmes
Content standards for digital geospatial data 338
Conversational Framework 227
cookie file authentication 370
Copernican view, of libraries 50
copyright
    copyright services in academic libraries 220
    potential conflict with rights management 226
    redefinition of for digital media 40
    role of librarians 252
    unauthorized use of digital material 73–4
    of web-based resources 161, 178–9
    see also intellectual property
core resources, identifying 224
Core values statement (IFLA) 245
Cornell University (US) 13
Corporate Yahoo! 167
Corporation for National Research Initiatives 314
cost effectiveness, of digital libraries 69–70, 79–82, 81